RE-ROOTING THE LEARNING SPACE

NEW DIRECTIONS IN MATHEMATICS AND SCIENCE EDUCATION
Volume 21

Scope

Mathematics and science education are in a state of change. Received models of teaching, curriculum, and researching in the two fields are adopting and developing new ways of thinking about how people of all ages know, learn, and develop. The recent literature in both fields includes contributions focusing on issues and using theoretical frames that were unthinkable a decade ago. For example, we see an increase in the use of conceptual and methodological tools from anthropology and semiotics to understand how different forms of knowledge are interconnected, how students learn, how textbooks are written, etcetera. Science and mathematics educators also have turned to issues such as identity and emotion as salient to the way in which people of all ages display and develop knowledge and skills. And they use dialectical or phenomenological approaches to answer ever arising questions about learning and development in science and mathematics.

The purpose of this series is to encourage the publication of books that are close to the cutting edge of both fields. The series aims at becoming a leader in providing refreshing and bold new work—rather than out-of-date reproductions of past states of the art—shaping both fields more than reproducing them, thereby closing the traditional gap that exists between journal articles and books in terms of their salience about what is new. The series is intended not only to foster books concerned with knowing, learning, and teaching in school but also with doing and learning mathematics and science across the whole lifespan (e.g., science in kindergarten; mathematics at work); and it is to be a vehicle for publishing books that fall between the two domains—such as when scientists learn about graphs and graphing as part of their work.

Re-Rooting the Learning Space

Minding Where Children's Mathematics Grow

By

Jennifer S. Thom
University of Victoria, Canada

SENSE PUBLISHERS
ROTTERDAM / BOSTON / TAIPEI

A C.I.P. record for this book is available from the Library of Congress.

ISBN 978-94-6091-428-7 (paperback)
ISBN 978-94-6091-429-4 (hardback)
ISBN 978-94-6091-430-0 (e-book)

Published by: Sense Publishers,
P.O. Box 21858, 3001 AW Rotterdam, The Netherlands
www.sensepublishers.com

This book has been reviewed by independent peer reviewers, who recommended publication.

Printed on acid-free paper

Dedication

To my Ba Chan, father, and mother, for situating me in their certain visions of the world.

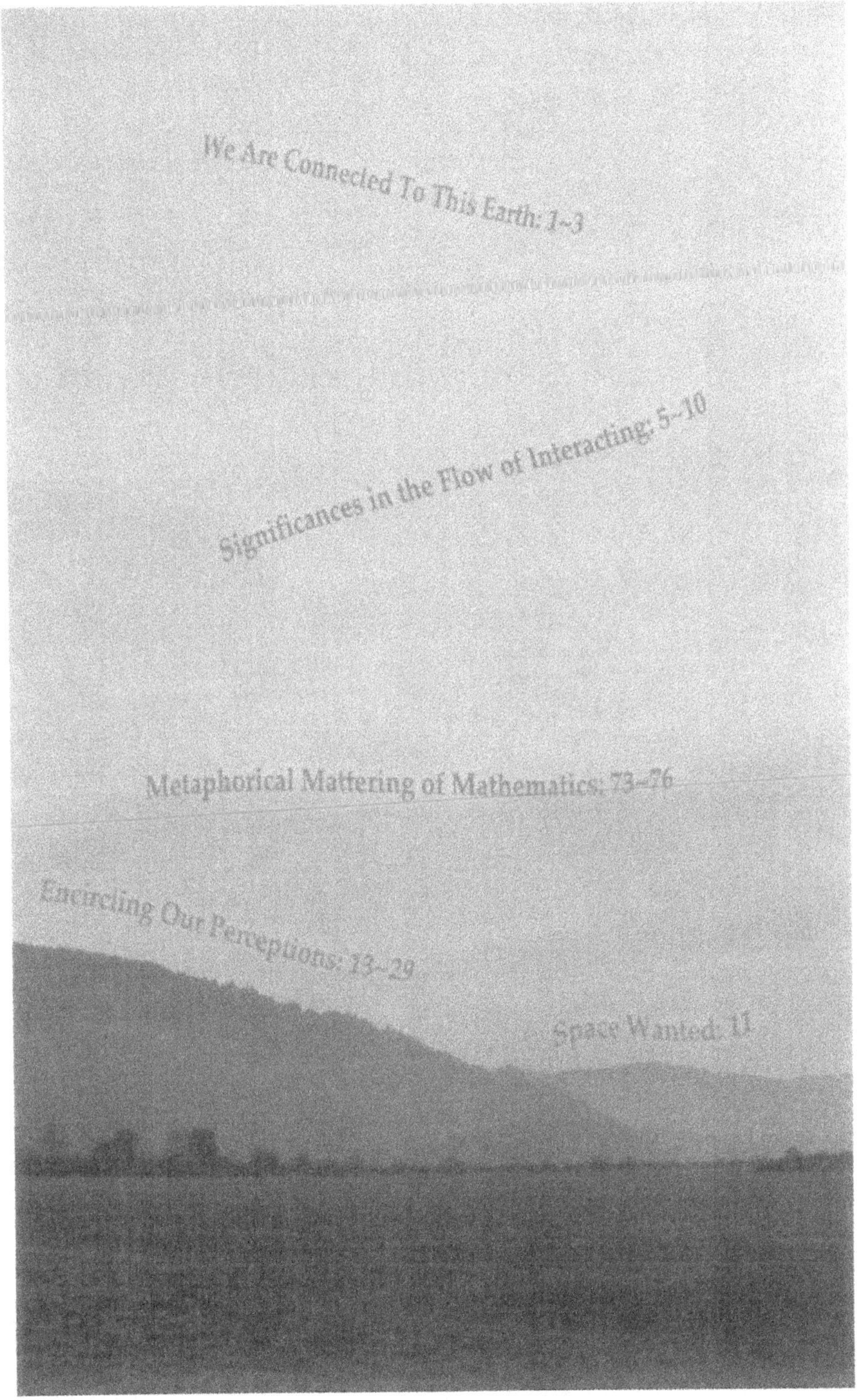

We Are Connected To This Earth: 1~3

Significances in the Flow of Interacting: 5~10

Metaphorical Mattering of Mathematics: 73~76

Encircling Our Perceptions: 13~29

Space Wanted: 11

Conceptualizations of Time: 181-198
Help Wanted: 31-44
New Furniture: 137-142
Traditions, Tensions, and Transitions: 45-72
Theoretical Portraits of
Mathematical Understanding: 77-135
Languaging Emergent Mathematics: 143-161
Atmosphere and Re-COGNIZING Differences: 163-171

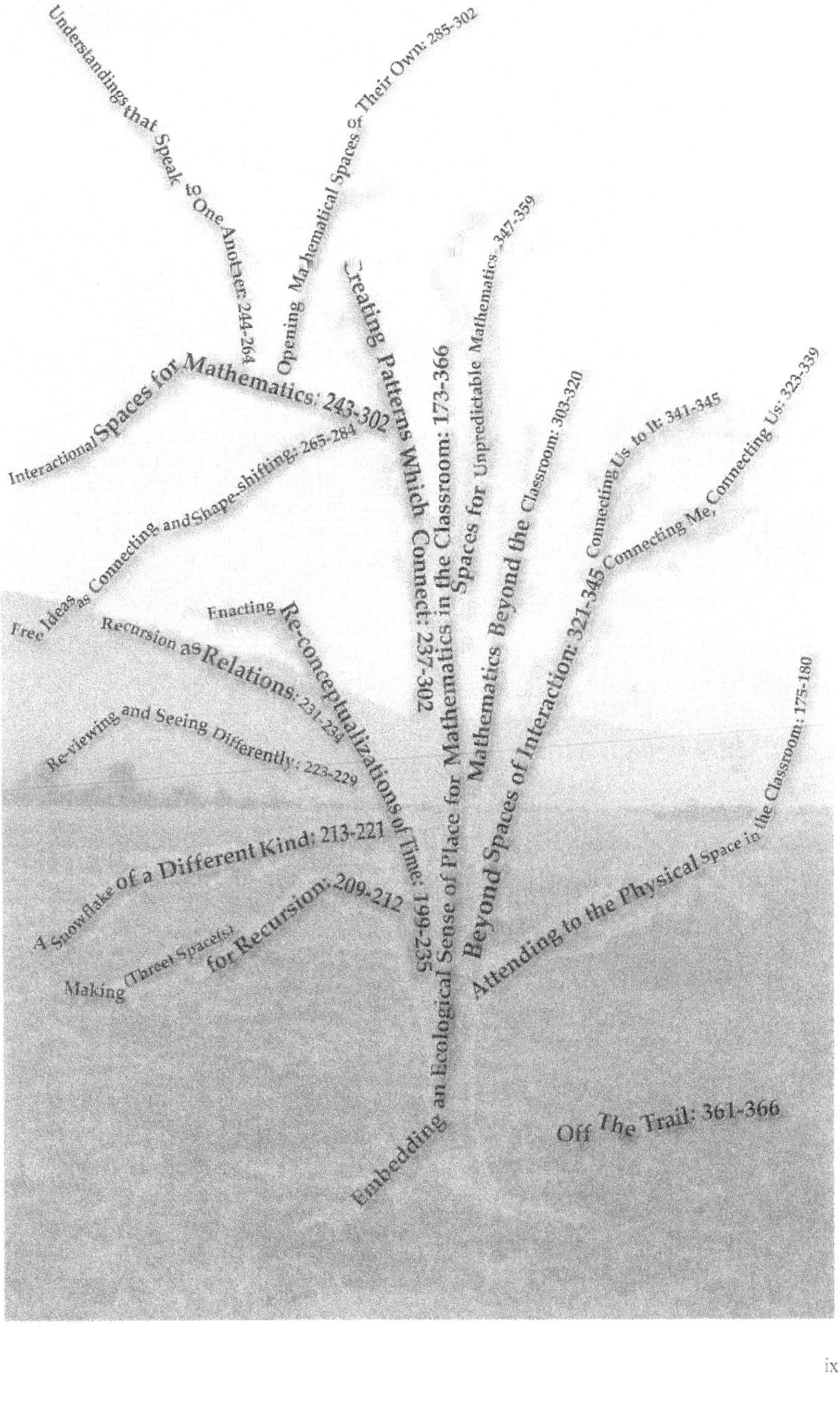
Understandings that Speak to One Another: 244-264
Opening Mathematical Spaces of Their Own: 285-302
Creating Patterns Which Connect: 237-302
Spaces for Unpredictable Mathematics: 347-359
Interactional Spaces for Mathematics: 243-302
Free Ideas as Connecting and Shape-shifting: 265-284
Mathematics Beyond the Classroom: 303-320
Connecting Us to It: 341-345
Connecting Me, Connecting Us: 323-339
Enacting Re-conceptualizations of Time: 199-235
Recursion as Relations: 231-234
Re-viewing and Seeing Differently: 223-229
Embedding an Ecological Sense of Place for Mathematics in the Classroom: 173-366
Beyond Spaces of Interaction: 321-345
A Snowflake of a Different Kind: 213-221
Making (Three) Space(s) for Recursion: 209-212
Attending to the Physical Space in the Classroom: 175-180
Off The Trail: 361-366

PREFACE

To understand a living system, such as a tree, in an ecologically systemic way involves more than simply reducing the tree down to its parts or by analyzing the tree from part to whole. Not only does one need to study the tree's leaves, stems, branches, trunk, root system, and its interaction with the environment but from many vantage points to make sense of how each part exists in dynamic relationship with the others as an integrated system.

The same is true about the purpose of this book. It is not meant to be a recipe for how to teach mathematics well or to serve as simply a descriptive account of a teaching practice. It is in essence, a systemic exploration into the embeddedness and co-emergence of theory and practice in mathematics teaching.

The nature of this inquiry and the theoretical realm in which it is situated necessitated the work to be expressed in a form that possessed an ecological sensibility. Upon first glance, this book appears to be a collection of separate compositions. Although each piece is an entity unto itself, the intent was not to render the book as a piecing together of theory, data, and analysis but instead, to bring a multiversal perspective to it and expose the co-existence and co-evolution of theory and practice. Thus, the embodiment of this book's thesis is also evident in the organization of the text as a whole and the diversity of writing structures within it.

The organic way in which this book is organized can be likened to a tree: that any one leaf is neither directly connected to the other leaves nor does one need to view them in any particular sequence. Yet at the same time, all are interconnected as essential parts of the tree by way of its stems and branches and therefore, are necessary for making the tree a coherent whole. These compositions which were sourced by video and audio taped sessions in the classroom, journal entries, students' work, and running field notes are not necessarily directly linked to one another or sequential in order. Each piece is considered to be a particular yet integral way of thinking that in turn forms larger conceptual clusters within an ecological mind-space. And together, these interrelated conceptual clusters as manners of knowing seek to in-form the mathematical learning space.

In the same way that one must explore a tree from many different vantage points in order to understand it, this book as an exploration of mathematics teaching also needed to possess different viewpoints. Moreover, it was critical that each piece of writing in some way highlighted the inevitably personal particularities that are involved in researching and teaching mathematics. To do this, different ***characters*** *were developed. The characters in this book are my students, a teaching colleague, and myself. In order to examine and interpret my teaching from multiple perspectives, my character took on several* ***personalities****. In some of the compositions, I am the main character and describe events as I perceived and experienced them either in the moment as they came to be or from a reflective stance after they had happened. In other instances, I do not exist "in person" or am another character altogether. These latter traits allowed me to interpret my examination from a connected yet more distant perspective. And in still other places in the book, the reader will find me in conversation with another character. These vignettes reveal the ongoing questioning and assessing of the theoretical coherence concerning my exploration.*

All of these different points of view are further communicated through the following expressive structures:

> ***Metaphors*** *and* ***visuals*** *are used to reflect how I was conceptualizing the theory and my teaching.*
>
> ***Featured quotes, ideas, or questions*** *positioned on a blank page are the theoretical artifacts that at the time, served as provocations for new inquiries. In this book, these quotes, ideas, or questions are intended to interrupt the reader and signal the opening of a new conceptual space.*
>
> ***Juxtaposition of text with other text*** *visually expresses the recursive and emergent manners in which my thinking arose during this study. Sometimes, juxtaposition was used to set ideas with each other so that I could discover what theoretical movement, friction, or generative spaces might be occasioned for my further consideration. While at other times, the juxtaposing of ideas enabled relationships to grow between one author's thinking and that of another's.*
>
> *The use of* ***black and white*** *or* ***shades of grey*** *for certain texts and images distinguish those theoretical underpinnings*

that I considered to be clearly defined and those I perceived to be fluid and indeterminate. For example, black on white is used in the text collage and portrait for constructivist notions to convey what I conceived as theory that was clear-cut whereas shades of grey were included in the ecological text collage and enactive portrait to express theory that I understood as imbued with qualities not so clear-cut but more unpredictable and ever-changing.

The actual **figuring** of the compositions — for example, as a newspaper article, conversation, or free form poem — as well as how each piece is organized in a left to right, top to bottom, back and forth, sporadic, or even whorling manner sought to capture the conceptual and metaphorical essence of the ideas being discussed.

Bolding of text is used to bring forth ideas in the conceptualization of theory while **fading of text is used to** preserve the contextual background from which these ideas emerged. This form of writing gives the reader a sense of the different visual fields as well as the depth of (those) fields that evolved in my thinking.

In addition to the characters, personalities, and writing styles, the actual **fonts** of the text distinguish between the different tones or perspectives taken on in my examination and interpretation of this work.

And lastly, although this book could be read in a conventional manner, I invite the reader to engage in the same spirit as one would exploring a living tree; perhaps examining the tree's integrated and integral being from its base and climbing up, swinging from one branch to another and gazing around, or even looking down at the always emerging tree and its surroundings from a distant hill.

Jennifer S. Thom
February 2012

ACKNOWLEDGMENTS

I owe a great debt of gratitude to many people who supported and contributed to the actualization of this book. I especially wish to acknowledge those not mentioned here who helped me during critical periods of the work.

I am extremely grateful to Susan Pirie who went beyond collegiality to encourage me to pursue this work. Her genuine interest in all aspects of the project, from the theoretical topics and practical instances of teaching mathematics to the actual styles in which these are expressed, made definitive contributions to the evolution of the book. The generative conversations we had occasioned so many insights that proved to be invaluable as I worked out both the conceptual and writerly details on various drafts of the manuscript. Just as integral was Wolff-Michael Roth's open invitation for me to explicate further, the concepts of the work and make it available to a larger audience. I have had the benefit of Tom Kieren's thoughtful readings and commentary on this work from its very inception. His theoretical prompts and intense, unwavering inquisitiveness were fundamental in the completion of this book; initially, in helping to define the direction that the work would take and later on, in articulating several of the ideas around which the book is structured. And without Karen Meyer, I wonder what my exploration into the discourse of ecology would have been like. Sincere thanks for setting me on my way and for helping me to find a place of resonance within it.

Ann Anderson's interest in the beginning stages of this work was important in furthering my understandings about classroom mathematics and my actual teaching of mathematics. Werner Liedtke and Chet Bowers, each in his own way, altered my perceptions about mathematics teaching and learning. On several occasions, their comments interrupted my thinking and caused me to re-examine my assumptions.

The conceptualization of particular ideas and compositions in the book also grew out of conference presentations and publications; both of which I received valuable feedback that was so important to the further writing of this work. *Our Cultural Ways of Being* includes ideas that evolved after and between presentations at Learning And The World We Want,[1] the University of Victoria,[2] and Narrative Matters,[3] as well as an article published in *Educational Studies.*[4] Parts of *Surfacing and Noticing* and *New Furniture* emerged from a theoretical paper published in *Mind, Body, and Society: Emerging Understandings of Knowing and Learning.*[5] As these writings continued to take shape, they were combined with empirical data of my mathematics teaching, presented at the annual meeting of the North American Chapter of the International Group for the

Psychology of Mathematics Education[6] and following the conference underwent further theoretical and interpretive revisions before assuming their current forms in the book. *Scripting an Unscripted Lesson, Mathematical Language, Languaging, & Residues of Learning, and Arisings* began as a presentation at the Complexity Science and Educational Research Conference.[7] Considerable alteration and conceptual development of these compositions was carried out for the book. *Mathematics Beyond the Classroom* includes some excerpts previously presented at the National Council of Teachers of Mathematics Regional Conference[8], the 2nd Meeting of Enactivism and Post-structuralism: Implications for Development of Educators[9], and an article published in *The Numerator.* [10] And, *Opening Spaces Of Their Own* started out as a presentation at the annual meeting of the American Educational Research Association[11] and later became an article published in *Journal of Mathematical Behavior.*[12] Like these other pieces of writing, it too experienced revisions, particularly in the conception of the narrative form and the visual layout that it now takes.

In addition, I wish to extend my immense thanks to others who have helped in various ways and especially, the Richmond School District for its dedication as an educational community. Not only was the district my educational home where I grew as a teacher but in a significant way, provided a rich environment for the co-emergence of my teaching and research to take place. During this time, it was the students who peaked my curiosity and compelled me to search for deeper understandings of what it meant to teach and learn mathematics. The teaching I did with Donna Jenner was truly a collaborative learning experience and indeed, a pleasure. I also want to express my appreciation for the financial support that I received from The University of British Columbia, the Richmond School District, and the Social Sciences and Humanities Research Council. These grants greatly aided my research and writing of this work. And, it was through Laurie Baxter's perceptive persistence and spirited, artistic ways, that the cover of the book emerged, visually embodying the meanings inside it.

Most importantly, I wish to thank my family for their kindness and guidance to see this work to fruition. In particular, gratitude goes to Pat for spending many a conversation helping me to craft my ideas and get them down on paper. I am especially grateful to Lucas for his care, patience, and constant encouragement. And finally, this journey would not have meant so much or been as wonder-filled without Nolan, Clara, and Myles.

A LETTER TO THE AUTHOR

Susan E. B. Pirie
France

February 7, 2012

Dear Jennifer,

You asked me to write something as an "introduction" or "afterword" to your book and this is a problem for me. In your preface you have said all that I would have said to introduce the book to the reader.

Anything I write would detract from your work and your thinking. In addition, it does not seem to me that an "afterword" is what is needed, since the piece is not to be read from cover to cover with a "and now what?" from me, to follow. I hope that the reader will hop about within the text as you recommend, will go back to what has been read with an insight from what has come later. It is not a book with a conclusion. There is no "end point" to understanding. It does not need further pointers to connections with other thinking because it is all about doing one's own thinking and path-laying or finding, isn't it?

This letter is both personal and yet public as is the book itself, which has taken many *years to reach the public domain. I feel I have accompanied your manuscript from its embryonic ideas to maturity. Yet in absolutely no sense am I its parent. You conceived it; you gave birth to it; you were its guardian through awakening youth, awkward adolescence to this momentous stage when it is ready to leave your hands and set out alone. A tough moment for a parent, saying "Fare well" to one's child, full of hope for its future success. Perhaps I can be considered its godmother, hovering over its cradle, following its progress, celebrating its achievements, always wishing mother and child a bright future. For the book will always be your "baby", just as will Nolan, Clara and Myles when they reach adulthood. You can never be sure you used all the right words at the right moments, gave sufficient thought to the actions that accompany a child's up-bringing. It is with trust that you must launch your progeny into the world. Have faith that the public will receive with curiosity and intellect the personal sides of Jennifer which they are offered.*

This is a book about growth: personal growth, pedagogical growth, mathematical growth and the growth of children's understanding. Metaphors have often played an important part in the communications between you and me and education is certainly in need of a new metaphor. The ecologically compatible approach of complexity theory encompassed by enactivism that you explore offers a powerful and exciting alternative to the existing, mechanistic, structural thinking underlying current curricula. It was Lakoff and Johnson's, Metaphors We Live By (1980) that first opened my eyes to the power of a metaphor as an analytical tool. Since then, much has been written by the mathematics education community around metaphor. For a period, the word itself created a bandwagon onto which many jumped using and abusing the notion in a variety of ways until its central power became dissipated and, at times, lost. We cannot avoid the fact that our existing concepts structure the way we interpret what we say and hear. Unaware, we are affected by the personal images and connections that language calls forth for each individual. We live our lives, albeit mostly unknowingly, immersed in a language of metaphor. Pause for a moment and reconsider part of my previous sentence: we live immersed in language. 'Immersed' calls upon the conceptual metaphor of an all-encompassing fluid, reflecting and influencing the smallest movement and thus the least expression of thought. Substitute the word 'surrounded' — we live surrounded by language — a metaphor of place, of physical situation, and immediately we have the feeling of a little freedom: others may talk at us but we have the choice of heeding or ignoring their implications. However I might have said "we are trapped by language" evoking a hostile environment within which we have no power;

we have no choice but to use words unaware of the images they may conjure up for others. For that is the tyranny and the power of metaphor, the image it evokes, the mental connections it engenders for the user and the receiver. Yet we can turn this power of the metaphor to influence thought to our own advantage, and use it as an analytical tool, a way to see differently, deliberately. Stay with my metaphor of fluids, of immersion in a language. What do we know about fluids that might help us to explore the use of language? What might one mean by “hot” language? “cold” language? “shallow“ language? ”the tide” of language? “rocks beneath the surface” of language? Do any of these new ideas of language invoke images that could shed light on how language is used? Perhaps we can play with these ideas sometime together!

Meanwhile if I were writing for your readers I would say, “pay close attention to the metaphors that Jennifer uses. Allow the images to arise that she conjures with her carefully chosen words. Use these images to see deeper into the complexity of understanding mathematics and children. Although highly theoretical in her thinking, her work is firmly grounded – she would probably say “rooted” – in her classroom. This book is no theoretical treatise, accessible only to academics. It is highly readable by the teachers she wishes to engage. Description, argument and interpretation are expressed through forms of portraiture, narrative writing and metaphorically enacted through layout and surprise. She offers vignettes which portray the co-emergence involved in creating an ecologically rooted sense of place for children's learning of mathematics, juxtaposed with diagrams and photographs which serve as visual images for her conceptual development of theory. These are no theoretically conceived teacher and children, but the author herself and her real students, with all the human quirks and imprecisions that constitute an everyday classroom. It is rare to find a work that combines, so effectively, theoretical originality and the reality of the classroom in such an enticing and readable form."

Since retiring to my small corner of French countryside, where fruit trees and vegetable plants, bees, chickens and lambs replace video recorders and students, my learning – because we never stop learning – has taken me down new exciting roads. My intellectual meanderings no longer intersect with yours directly but it gives me enormous pleasure to be able to watch the launching of your book because I see you moving off under your own impetus; taking our thinking together and that of others to new realms. You are laying down your own path and need no guide. The thoughts, ideas, diagrams, etc., I have left behind are there for you to carry, explore, throw away as you see fit on your journey. Do what you will with them.

All I feel entitled to say to your readers is, "once I was fortunate enough to have her as my student and now she is journeying alone, carving paths and planting trees through the wilderness of learning. Take her hand and travel with her. That way you will find your own new and challenging paths to understanding."

What I really want to say to you, Jennifer, is, "Thank you for enriching my thinking and ways of being during our time together. Your readers have a treat in store."

Very best wishes,
Susan

WE ARE CONNECTED TO THIS EARTH

As we continue to pour chemical cocktails into the environment and move fast and furiously from one technological adventure to another, it is no longer a matter of choice but a matter of fact that in order for living systems to survive on the earth, we must live within its limits of sustainability.

Upon our clumsy awakening to the environmental crisis, we are presented with the realization that the impact of our actions cannot be contained and the effects of them reach farther and deeper than we ever anticipated. The ongoing devastation of the world's natural and cultural systems makes this point clear: The results of how we live are not only felt by the local human community and our neighboring communities, but what we do affects all that is on this earth with us – the land, water, air, and every living being that depends on these sources for their existence. We are not independent beings. We are "just one particular strand in the web of life," says Capra.[1]

Yes, but we recycle!

Deep changes cannot happen if our actions to reduce pollution and decrease stress on the earth's natural systems remain entrenched in our desires to improve human health and maximize profit at the cost of all other forms of nature. If we are to prevent further damage to the environment it is imperative that we change our mechanistic perceptions of the world to ones that are ecological.

Simply put, this means abandoning the reproduction of our Cartesian-self-assertive-Newtonian manners of being in order to cultivate a more integrative existence on this earth. It entails re-rooting our thinking so that we may comprehend the world in nonlinear, connected, intuitive, and holistic ways; ways which contrast with those that are linear, analytical, rational, or reductionist in approach. Furthermore, value needs to be placed on cooperation, quality, and conservation instead of anthropocentric, exploitive, or competitive acts of domination and mindsets that focus on 'the bottom dollar'. It is likely that these changes will involve more than just exchanging our current ways of living for ecological ones. The mechanistic, anthropocentric traditions that we embody in our culture today have been evolving since the Industrial Revolution and so too will it take time for eco-centric practices to become taken for granted patterns in our thinking, actions, and identities.

For eco-centric thinking to bring about changes that possess depth, longevity, and integrity, it cannot be restricted to environmental cleanliness. In fact, eco-centric thinking must become our natural, everyday way of being. Further still, learning to be ecologically mindful cannot be treated as an additional component in students' education but has to be an integral part of each and every classroom. In short, this means that teachers and students of mathematics may not be excused.

QUESTION:
So, what does classroom mathematics have to do with this?

ANSWER:
Everything!

Okay, but where does one start?

One place to start could be to look for spaces in which to propagate nonlinear, connected, fluid, and holistic patterns of thinking mathematically. At the same time, educators could begin the process of assessing the embedded and taken for granted linear and mechanistic rituals practiced inside the classroom and within the larger field of mathematics education.

By not excusing ourselves from this task, we can work towards reconnecting and developing ecological ways of teaching and learning mathematics.

What Is A Tree?

A tree, we might say, is not so much a thing as a rhythm of exchange, or perhaps a centre of organizational forces. Transpiration induces the upward flow of water and dissolved materials, facilitating an inflow from the soil. If we were aware of this rather than the appearance of a tree-form, we might regard the tree as a centre of a force-field to which water is drawn. The object to which we attach significance is the configuration of the forces necessary to being a tree.... It is not necessarily the case that whatever exists must be sharply bounded; in fact, rigid attention to boundaries can obscure the act of being itself.

- Neil Evernden, The Natural Alien[2]

This redefinition of something as familiar as a tree at first rings strange. But we can recognize the more-than-tree-form it describes, just as we know that a forest is more than just the trees that grow there, and that our intercourse with the world extends beyond the edges of our skin. Our language falls short of our apprehension because of the way we have been taught to identify the world. We belong to, are made of, that world that surrounds us, and we respond to it in ways beyond knowing.[3]

(continued from page 24)

Today we are calling on the nations and the peoples of the world to change personal attitudes and practices.[4]

"To adopt the ethic for living sustainably, people must reexamine their values and alter their behavior. Society must promote values that support this ethic and discourage those that are incompatible with the sustainable way of life" (from *Caring for the Earth: A Struggle for Sustainable Living*. Gland, Switzerland: IUCN/UNEP/WWWF, 1991)[5]

(continued on page 30)

Reflection

A mirror reflects an image
seen as the image is seen.
It does not change the
looking.

To reflect on what we do,
or are, is something else.
It reveals what we could not
see.

~ Pille Bunnell[1]

We are constantly engaged in the flow of interacting but often, it is not until much later that we appreciate the significance of it.

Significances in the Flow of Interacting

It happened years ago. Jennifer could not remember exactly when but as a young child she was drawn into the enchantment of the "enveloping and sensuous earth".[2] Eyes wide and bright, Jennifer stories out just some of her treasures to me. She speaks of wild landscapes just beyond her grandparents' orchard; tunneling on her stomach and disappearing into the tall, sweet grass; lying on her back and watching the night sky for cascading meteorites; and feeling the cool dampness underfoot as she crept through the grass and listened hard in search of that mysterious chirping cricket. Among these treasures are many more: ones of forests and of the ocean, others storied to her by her Ba Chan's tanka poems, as well as the Chinese proverbs told to her by her father.

> "I suppose," said Jennifer, "because my family life has always been rooted in a kind of living that looks to nature for metaphors and life lessons, I also want to understand the world as a unity in which everything is interrelated. And I guess that's why I wonder how ecological forms of thinking might help me to better understand this place we call the mathematics classroom."
>
> "Classroom mathematics and ecology? Interrelated? Tell me you're not serious," I said.
>
> "I know, I know," she said. "But listen."

Anticipating a long, winding, twisting, turning kind of response, I straightened my posture and took a deep breath as Jennifer led me down her explanatory path.

> "You see," she said, "ecology and classroom mathematics have everything to do with each other."

Jennifer told me that the word, *ecology*, came from the Greek word, *oikos*. It meant the family household and the maintaining of its daily operations. Eventually, *oikos* was integrated into the term, *oecologie*, which was coined by Ernst Haeckel in 1866. Described as the study of the environmental conditions of existence, oecologie was eventually shortened to what we know today as *ecology*.

> "I remember exactly what I was doing when I first learned about the history of the word," said Jennifer. "To be quite honest, even though I'd heard what had been described, I didn't give it much thought at the time because I was preoccupied with what I assumed to be a completely different matter. You see, I was taking a summer graduate course and we were discussing the origin of the word. I remember smiling and nodding as I listened only to return to my

previous thought. It wasn't until much later that I realized the significance of the encounter."

Jennifer explained that the dilemma preoccupying her thoughts concerned her awkward attempts to communicate to others, the importance of being ecologically mindful in the mathematics classroom.

"You see, at the time it was easier for me to say what being ecologically mindful didn't mean. Being ecologically mindful wasn't necessarily about taking environmental issues and turning them into mathematical problems. It wasn't about conducting scientific investigations... you know, taking ecological ways of thinking and using them as a magnifying lens to examine issues in mathematics education and then perform experiments in the classroom; that is, it wasn't about forming a hypothesis, replicating procedures, generating conclusions or formulating a unifying theory that could be transplanted into every classroom. What I struggled with was that I didn't have a simple explanation of what being ecological did mean for the mathematics classroom. My descriptions were cumbersome – that knowing and acting were embodied within one's way of being – or, a mindful comprehension of the integrative, holistic, and nonlinear nature of teacher's practices and students' learning of mathematics. So while I spent my time trying to sort out the ideas that I viewed to be critical in developing an ecological sensibility for teaching and learning mathematics, I was completely ignorant of the fact that I did have a way to express my understanding of ecology and mathematics education. Oikos," said Jennifer.

Jennifer told me that upon reflection, the term, oikos, captured exactly how she understood her mathematics class to be. For her, she imagined the mathematics classroom to be very much like a family household. As teacher, she saw her role as caring for and sustaining the mathematical relationships and interactions of her students.

"So just as ecological thinking about the environment focuses on human relationships with nature," Jennifer smiled, "I have a similar focus for my teaching and my students' learning of mathematics. As the students' teacher, my focus is on examining and assessing the kinds of mathematical relationships as well as the forms of mathematics that emerge in the classroom, and responding to them in my manners of teaching mathematics."

"And your reason for wanting to be ecologically mindful?"

> "My wanting to be an ecologically responsive mathematics teacher comes from caring about how mathematics exists in the classroom, my teaching, and the students' learning of it. This kind of caring necessarily involves a commitment to nurturing relations that are not only ecologically coherent in the classroom but also ones that promote a sense of coherence within the larger educational communities."

Jennifer picked up a book that she had been reading before I arrived. Opening it to page 78, a composition written by M. C. Bateson, she read aloud:

> We are living in a time of both creativity and concern about education, and the decisions that are made for the classroom will feed directly into the way graduates

"and students," she added,

> participate in society and the way they impact on the natural

"and social-cultural," she said,

> systems around them.[3]

"And so you see," Jennifer said, "the choices we make as mathematics teachers not only affect the kinds of mathematics students learn in school but more profoundly, the ways in which they are *taught* to learn and the ways in which they interact with mathematics outside of school affect the world they live in. We, mathematics education, and ecology do not exist in separate households but rather, we share a common space."

SPACE WANTED

Looking to share a space with ecology. Interested in what ecologically coherent forms of teaching and learning mathematics could mean for the classroom. Can move in IMMEDIATELY.

HELP WANTED

"How can we break out of our conventional approaches and imagne more productive alternatives?"[1] Reply to mailbox: T1I9M9M7S

VACANCY

Seeking one teacher to teach grades 2/3. Separate room. Shared facilities.

(continued from page 36)

3 Faces of Ecology

M. C. Bateson[2] identifies three "faces" or realms of ecology: empirical, enviromental, and systemic. The author defines empirical ecology as biological, meteorological, and geographical studies that focus on understanding how the planet is changing and how these changes affect the interrelationships of the world's natural systems. The environmental face of ecology is concerned with identifying the level of impact that our ways of living have on the earth's systems. This area of ecology also involves the development of solutions for environmental problems that minimize harmful stress on the earth. It is within the systemic realm of ecology where mathematics teaching and learning can be most radically explored. This is because systemic thinking focuses on seeking "the pattern which connects"[3] a system or systems together as interacting wholes.

A "system" could be a teacher or a student. It could also be a collective group such as a mathematics class, a school, and so on.

The connecting pattern or patterns that interrelate one system to another encompass the forms of knowledge, actions, and identities that result from the ongoing interactions within the systems as well as the ways in which these dynamics are sustained by the systems themselves.

Focusing on relational qualities makes it possible to view the world as an integrated whole; a dynamic and fluid unity in which all living and social-cultural systems are interconnected. The world is not conceptualized as a collection of separate entities but instead, as a highly complex unity in which all systems are interrelated and therefore, interdependent.

It seems sensible that when looking from an ecological perspective, mathematics teaching could be conceived as similar to that of students' mathematical learning.[4] Mathematics teaching then, as a fluid and complex process, implies that it exists always in relation to the ongoing interactions of the students, the mathematics, and the material and nonmaterial environment of the classroom.

In light of this, it is from a systemic perspective that the following question is posed:

> ***In what ways can systemic manners of thinking about mathematics education enable forms of teaching and spaces for learning mathematics to possess an ecological sensibility?***

Yes, but what gives rise to a systemic, ecological view of the world? or the mathematics classroom for that matter?

Our perceptions of the world and the mathematics classroom...

Our individual ways of being...

I think that if we start by looking at how we are as individuals and then connect this to how we exist as collective groups, you'll be able to appreciate why a systemic understanding of the world and the mathematics classroom is really about the process of living. This kind of understanding defines a conceptual space in which we can make visible what often remains invisible — the co-emergent, complex nature of our

B biological,

E structure determined,

I social,

N cultural

G

Okay, let's begin.

Well, as humans, we exist in the world as what Maturana and Varela would call, autopoietic, or self-making systems. What this means is that we possess both organization and structure. It's our organization that distinguishes you and me as people and not, say, fish or goats! And it's our structures — our internal dynamics and relations which enable you and me to develop ways of knowing, acting, and being that are uniquely our own. Simply put, your structure isn't the same as my structure and it's because of our structural diversity that we can distinguish you and me as being different people.

But how is it that we are structurally different?

Maturana and Varela describe structural coupling as the process by which our structures evolve. The changes that occur in our knowing, actions, and identities arise from the recursive interactions between two or more living organisms.

A nice sounding definition but what does this mean?

Well, if we take this idea of structure and think of a person's understanding of mathematics to be his or her mathematical structure, in a way similar to how a person's forms of knowing, acting, and being are impacted by life experiences — a person's mathematical structure too undergoes changes as a result of his or her mathematical interactions.

And so, because one's structure is dependent on the kinds of mathematical interactions one has and how they then feed into what the person already understands, these differences in experience and impact create differences between one person's mathematical understanding and that of another person's... therefore, diversity in mathematical structures.

Yes. And recursively, how we go on then, to teach or learn mathematics will be shaped by these structural differences. This also means that as a class engages in mathematics, structural coupling may arise in the structures of the individual students and their teacher. The growth that results from this process is dynamic and continuous. It happens in us moment to moment as we experience what you could call, human and nonhuman perturbations in the environment.

So it's the perturbations that make for structural changes?

No, not exactly. It isn't the perturbation that determines how our structures change. And, perturbations only exist if the person perceives these as perturbatory. Rather, it's the individual based on his or her structure, who specifies what will or won't be a perturbation, whether or not coupling will occur, and if so, the kind of internal changes that will arise in his or her structure. Knowing this, we can say that we exist in the world as autopoietic and structurally determined systems.

Would this mean then, that in the mathematics classroom, it's the child who determines based on his or her internal structure, what will and will not serve as occasions for learning mathematics?

Yes, and the child's mathematical understandings — his or her structure, shapes and is shaped by future understandings.

But what about the teacher? Isn't it the teacher who teaches the class what mathematics to learn?

Of course it should be expected that a teacher attend to children's mathematical learning in ways that are invocative[1] and provocative.[2] However, with a systemic view, we can't assume that the teacher exists as the only source for occasioning children's mathematical perturbations. Engaging in mental reflection about mathematics or taking part in mathematical interactions with the human and nonhuman environment also serve as possible sources for structural changes to occur.

So, even if a teacher intends to have the class learn, say, a new strategy for adding three-digit numbers together, it's the child, not the teacher who determines if and what kind of learning will arise?

Exactly. And when structural changes do take place, new pathways or relationships emerge and impact the child's existing mathematical understandings. So it's impossible for us to predetermine how our individual structures will evolve since they are ever-changing because of the coupling process. This is what Davis, Sumara & Luce-Kapler mean when they say that learning is dependent on, but cannot be determined by teaching.[3] Mathematical learning takes place with the environment: as unpredictable yet recursive growth of one's mathematical structure of understandings.

Okay, I can see how we as individuals are shaped by the interactions we have with the environment but it seems to me to be a very inward, insular way to view the mathematics classroom, don't you think? This kind of thinking moves in only one direction — from the environment to the individual child.

And up to this point it has. However, a systemic view does bring forth a co-emergent worldview, if you will, in that it recognizes the interdependence and complex circularity that exists between the environment and living systems. Just as our internal structures are ever evolving through our interactions with the environment, the environment is also undergoing structural changes. These changes with-and-in us and the larger environment recursively shape what's possible in terms of future interactions and how each responds to the other. Lewontin elaborates on this complex circularity when he explains that:

> The organism and the environment are not actually separately determined. The environment is not a structure imposed on living beings from the outside but is in fact a creation of those beings. The

environment is not an autonomous process but a reflection of the biology of the species. Just as there is no organism without an environment, so there is no environment without an organism.[4]

This co-evolution that takes place as the environment and we interact raises an important issue when considering the mathematics classroom. It's not only the environment that shapes the teacher's or a child's mathematical ways of knowing, acting, and being, but it's also the teacher's or child's interactions that affect what future events and responses take place within the larger classroom.

Each needs the other.

That's right. Now can you begin to see how it's possible for the world to be viewed as an integrated whole by recognizing the interdependence of living systems and their environments? Life arises by way of natural drift — as a result of the recursive interactions between living systems and the environment. This co-evolutionary view of the world differs from other perspectives that project images of evolution as being a linear process of competitive domination whereby species and their environments are not interdependent but separate from each other.

Yes. A subtle yet important difference, I suspect.

What's more, a systemic, ecological view doesn't portray mathematics teaching or children's mathematical learning as being individualistic and linear in nature. Each arises fluidly in relation to the other and with that of the larger environment be it a mathematics class, a school, or even the educational system. An ecological perspective brings attention to understanding the interrelationships that exist within the mathematics classroom.

In the beginning of our conversation, you mentioned that we are also social and cultural beings, yes?

Yes, that's right. Keeping in mind what we've discussed in terms of how we are as individuals and how the environment and we co-emerge, let's move outwards to the broader, social realm. By doing so, we can continue to discuss the pattern which connects[5] *our living as individuals to our collective actions, identities, and ways of knowing as social systems.*

Our social ways of being…

Maturana characterizes social interaction as being:

> ... when two or more structure determined systems interact recurrently with each other in a particular medium, they enter in a history of congruent structural changes that follows a course that arises moment after moment contingent on their recurrent interactions, to their own internal structural dynamics, and to their interactions with the medium, and which lasts until... they separate. In daily life, such a course of structural change in a system contingent on the sequence of its interactions in the medium in which it conserves organization and adaptation is called 'drift'.[6]

In terms of the classroom, this would mean that social mathematical interactions arise when two or more children work mathematically together. Importantly, learning occasioned from these mathematical interactions not only shapes the further development of each child's understandings but also, the collective understandings of the partner or group and the larger mathematical environment in which the interaction took place. These collective forms of knowledge, actions, and identities and how they're created through social interactions are what Maturana refers to as drift.

If we understand human social systems to be what Gregory Bateson and Maturana refer to as systems that evolve through the cohesive, collective interactions of the members, then what we know, how we act, and who we are can't be taken as happening only within the realm of the individual. Such growth also needs to be recognized as emerging from our collective manners of living — the relations that are created through ongoing interactions and that which connects us as interdependent, social beings.

And are social phenomena unpredictable like our individual structures?

Yes. Just as we can't predetermine the evolution of an organism or its environment because they are dynamically interactive, we can't predetermine the collective mathematical activity that will take place in a mathematics classroom. In terms of a class' mathematical learning, it's naturally unpredictable because children's internal and collective structures are constantly changing from moment to moment.

I see. So a child isn't only a structurally determined learner but he or she is also a member of larger social systems... such as a mathematics class?

Exactly. And in the classroom it's not only the child but also the children and their teacher interacting at the same time... as individual and collective wholes, responding to environmental perturbations — shaping and being shaped by the mathematical learning that emerges. Like our individual understandings, collective forms of learning aren't thought of as being transmitted from an external entity. They are constantly emerging and co-emerging through social interactions. Because of this, children's mathematical growth can be described as being, much like paths that exist only as they are laid down in walking.[7]

Can you explain in more detail, the nature of social interactions and what forms they might take on?

In terms of their nature, I think of these as Maturana does — like conversations in progress. Maturana explains that social interactions can be brief, withdrawn from and then re-entered again or, they can be continuous. It is these conversations within social systems that he considers necessary in how it is we come to know and be in the world. Social dynamics are what bind us as a pair or group of living beings together as a collective, social system. Co-emergence takes place as we interact with and in relation to one another... we are able to coordinate and re-coordinate how we think, our actions... basically, how we are, in order to maintain cohesive ways of being with one another. In this way, social relationships that keep a collective unity intact can be seen as similar and just as critical to the co-evolution that takes place with individual organisms and their environments.

Maturana's idea of languaging is useful because it describes the process by which social systems function and evolve as collective unities. Now, it's important that you don't think of languaging as simply individuals engaged in verbal conversation with themselves or others. Languaging involves the physical, verbal, and mental ways we humans think and interact among one another, but it's also the understandings that arise from such linguistic interactions. It's how we are able to coordinate and re-coordinate our ways of being so that we can continue to interact within groups and develop collective forms of knowing. In other words, languaging in the mathematics classroom entails the mathematical understandings that emerge from the different ways in which members of the class think and engage mathematically with one another. Because we exist in language and have the potential to be languaging agents, it's possible for new understandings to arise. Knowledge systems evolve then, as a result of our social activity.

So does this mean that it's through the interactions of the class that collective mathematical understandings that are different from personal ones come into being?

Yes. Now, let's talk about our cultural ways of being.

Our cultural ways of being...

The practices of teaching and learning school mathematics are two examples of cultural behaviours in our society. Generally speaking, cultural behaviours are social patterns that span generations. It's as a result of our living in the languaging process of cultural drift that we establish collective ways of being that are passed on and evolve from generation to generation.

So cultural ways of being are social phenomena that continue over generations?

Yes. And the historical transformation that happens is a result of the recurrent interactions and languaging between the older and younger members of a cultural group. In the same manner that drift is explained by Maturana and Varela as necessary for us to evolve with the environment and socially with others, so too is cultural drift necessary for the continuity and evolution of cultural systems.

Okay. I see how cultural drift provides a systemic way to understand, say, how human-centred and mechanistic social patterns established in the Industrial Revolution have continued into today's culture. But what I don't yet understand is what ecological thinkers such as Bowers, Capra, Naess, and Orr mean when they say that our cultural ways of being shape how we perceive and therefore, live in the world.

We do much more than simply exist in the world — remember our conversation about coupling?

Yes.

Well, I believe that connected to this understanding is the fact that **THE WORLD(S)** *we bring forth isn't only created through our individual and social existence. It's also* **SPECIFIED** *through our cultural knowledge, actions,* **AND** *identities. Cultural phenomena, when* **CONSERVED**, *are carried forth seamlessly from one generation to the next and create a world from our living* **WITHIN CULTURAL COGNITIVE CIRCLES.** *These cognitive circles arise from our cultural ways of living and we justify the patterns that they occasion as a matter* **OF TRADITION OR**, *less reflectively and more acceptingly expressed, we say it's simply* **JUST THE WAY THINGS ARE.** *In this way,* **OUR CULTURAL PATTERNS** *embedded in our thinking* **ARE US**, DISAPPEARING FROM THE SURFACE OF OUR CONSCIOUSNESS.

Surely, our cultural actions, beliefs, and identities aren't that invisibly specific!

Hmm… well, consider the images we as members of Western culture attach to the idea of what it means for a person to be an individual. When we think about what makes a person an individual, often embedded within this is the notion of independence. As a teacher, I find that parents often express to me that it's important for their children as individuals, to be self-sufficient, able to think for themselves, make independent decisions, and be their own people. In valuing these qualities, we teachers and members of older generations encourage younger generations to develop their independence by providing learning opportunities that focus on the individual or autonomous child. Within our culture, independence and individuality serve as distinguishing qualities of being successful. They engender a sense of freedom, self-reliance and standing out from the crowd.

Well, isn't that what we should be doing? Encouraging students to become independent individuals?

Hold on for a moment. Let's contrast this image with what it means to be an individual in Japanese and Chinese cultures. Traditionally, within these two cultural circles, the image of an independent individual is not the image that comes to one's mind. This is because in Japanese and Chinese cultures the older members teach the younger members that an individual is not defined in terms of self-reliance or self-sufficiency but more in how the individual contributes to the wellbeing of his or her family. You see, a person's identity exists in the collective sense of the family. This can be seen in how people address one another. Unlike in Western society where we are distinguished on a first name basis such as Jennifer or in a first-name-last-name order as Jennifer Thom, people in Japanese and Chinese cultures are addressed by their last name or in a last-name-first-name order such as Thom-san or Thom-Jennifer. Identity isn't derived from the validation of one's self, but from the respect for the family as a collective whole. Conservation of these relations is carried forth through one's values and action that foster the wellbeing of the family as a collective whole. Given instances such as these, we can better understand what Maturana speaks of when he says that it's in the implicit or explicit accepted premises under which their different kinds of discourse, actions, and justification for actions take place[8]… that cultures create taken for granted and, hence, invisible yet distinctive cognitive circles.

Okay, yes… how cultural beliefs create blinders that shape how we experience the world… but, if we're truly blind to our cultural ways of being, is it even possible for us to become aware of them?

Another way for us to examine just how culturally embedded our lives are is to consider the cultural experiences, beliefs, and values that emerge from recurrent interactions of a group and metaphorically become what Gregory Bateson and Bowers refer to as a culture's map.

A map?!

Yes, a map. Simply put, The term and concept borrowed from Korzybski means that a culture's map identifies what its members will and won't perceive as having significance by rooting these features within the culture's temporal, spatial, spoken, written, and symbolic language. Because these cultural distinctions permeate the group's languaging, a culture's map as Postman says, does much more than construct concepts about the events and things in the world; it tells us what sorts of concepts we ought to construct.[9]

Can you give me an example of a feature or concept that we create or recreate in our living out of these metaphorical maps?

Just look back at how the idea of the individual is distinguished and played out as a feature of Western and Asian cultural maps. The former very much influences a person to value ways of thinking and actions that enable an identity of independence while the latter, emphasizes a person's connection to her or his family and imbues a sense of interdependence. So, they both share the same concept — the individual, but are two completely different cultural conceptualizations.

Yes, which lead to totally different ways of being in the world.

Speaking from a systemic, ecological thinking space, it's here in the envelopment of the cultural realm... nested within our individual and collective ways of knowing and being that we dwell in our practices of teaching and learning mathematics. For me, the mathematics classroom is imagined as an integrated space where individual, social, and cultural systems co-exist and co-evolve. In an ontological manner then, we live simultaneously as biological, structure-determined, social, and cultural beings. Encircled once more to include all other living beings and natural systems on the earth, it's how we humans come to exist, as Capra says, as just one particular strand in the web of life.[10] *It's in this way that the world isn't perceived as a collection of separate parts but as a dynamic whole; a complex unity of all living and social-cultural systems that are fluidly interconnected and, therefore, necessarily interdependent. And it is here in this conceptual space that a systemically ecological view of the world and the mathematics classroom resides.*

All living systems within the biosphere...

Our cultural ways of being...

Our social ways of being...

Our individual ways of being...

SPACE WANTED

Looking to share a space with ecology. Interested in what ecologically coherent forms of teaching and learning mathematics could mean for the classroom. Can move in IMMEDIATELY.

HELP WANTED

"How can we break out of our conventional approaches and imagine more productive alternatives?"[1] Reply to mailbox: T1I9M9M7S

VACANCY

Seeking one teacher to teach grades 2/3. Separate room. Shared facilities.

(continued from page 36)

3 Faces of Ecology

M. C. Bateson[2] identifies three "faces" or realms of ecology: empirical, enviromental, and systemic. The author defines empirical ecology as biological, meteorological, and geographical studies that focus on understanding how the planet is changing and how these changes affect the interrelationships of the world's natural systems. The environmental face of ecology is concerned with identifying the level of impact that our ways of living have on the earth's systems. This area of ecology also involves the development of solutions for environmental problems that minimize harmful stress on the earth. It is within the systemic realm of ecology where mathematics teaching and learning can be most radically explored. This is because systemic thinking focuses on seeking "the pattern which connects"[3] a system or systems together as interacting wholes.

A "system" could be a teacher or a student. It could also be a collective group such as a mathematics class, a school, and so on.

The connecting pattern or patterns that interrelate one system to another encompass the forms of knowledge, actions, and identities that result from the ongoing interactions within the systems as well as the ways in which these dynamics are sustained by the systems themselves.

Focusing on relational qualities makes it possible to view the world as an integrated whole; a dynamic and fluid unity in which all living and social-cultural systems are interconnected. The world is not conceptualized as a collection of separate entities but instead, as a highly complex unity in which all systems are interrelated and therefore, interdependent.

It seems sensible that when looking from an ecological perspective, mathematics teaching could be conceived as similar to that of students' mathematical learning.[4] Mathematics teaching then, as a fluid and complex process, implies that it exists always in relation to the ongoing interactions of the students, the mathematics, and the material and nonmaterial environment of the classroom.

In light of this, it is from a systemic perspective that the following question is posed:

> ***In what ways can systemic manners of thinking about mathematics education enable forms of teaching and spaces for learning mathematics to possess an ecological sensibility?***

Settling In

Thinking about how I (it's me, Jennifer) might respond to this help wanted ad, I thought it best to bring it home as it were and invite Stigler and Hiebert's question into this ecological thinking space of mine. However, once inside, I soon realized that although this question certainly belonged to the realm of classroom mathematics, it was going to be difficult if not impossible for me to have an open conversation with it. Explained another way, it is like when you spot THE sofa in a furniture store but as soon as you get it home and put it into your living room, the sofa does not look so fabulous anymore. Instead, it is clearly out of place because it does not go with any of your existing furniture. For me, (and in spite of the authors' good intentions) this seemed to be the case with bringing this question home; neither the question nor the ecological space suited each other. You see, the manner in which the question is posed:

> How can we **break out of** our **conventional approaches** and imagine **more productive** alternatives? [5]

puts forth for me as a teacher, an **'end-result'** mindset of **improving productivity** in the mathematics classroom; the need to **dispose of** or **discard** teaching practices that are perceived to be **old** or **commonplace** and to **acquire new** teaching **tools** so that we may **increase** or **maximize** children's learning of mathematics.

Before I am able to think about possible ways to consider and respond to this question, its linear, disconnecting posture has already told me that manners of teaching are commoditizable relationships, ones to which we essentially become engaged, eventually marry, and if necessary, divorce ourselves from. What is more, is that within the question's reductionist confines, Stigler and Hiebert's query closes itself off from the opportunity for deep changes

to

take

root.

You see, even if we changed from one teaching style to another, radical shifts in the mathematics classroom would not likely occur if our thinking continued to be fashioned from the mechanistic pattern of productivity. The persistence of such a mindset is disabling in that it denies the very possibility of Stigler and Hiebert's query to be one that provokes teachers in becoming more integrative and creative.

And because the authors' question does not allow for an examination of the *ecological coherence* of the mathematics classroom, it creates the impossibility for mathematics teaching and learning to be conceived as *holistic, organic, recursive, and co-emergent.*

Determined to re-conceptualize Stigler and Hiebert's query within an ecological realm, I set out to find a question that would make sense in such a space. This question had to be one that encouraged a systemic way of knowing: one that necessitated the development of an understanding for the ways in which our processes of living[6] are always and necessarily shaping how we teach mathematics.

Instead of provoking a knee-jerk reaction from the reader or myself, the question needed to be an open invitation to look deep and make visible what often remains invisible – the cognitive circles and cultural maps[7] – we embody and enact as we teach. Once visible, these experiences, beliefs, and values can be examined and assessed for the kinds of mathematics teaching that they enable and disable, the ways in which they become embedded within the language of the classroom, and the impact they have on how children come to know mathematics. While I wondered about these points, a similar yet radically new question invited itself into my thinking:

> *How might we as teachers consider our patterns of mathematics teaching? And, if necessary, how might we re-root the learning space so as to nurture and children's mathematical growth?*

As I reflected on this question, I realized that not only does the question share Stigler and Hiebert's concern for how mathematics teaching and learning takes place in the classroom, but it also makes sense within a systemically ecological thinking space. The question provides the necessary structure for the re-conceptualization of mathematics teaching to occur while the ecological mind-space allows a place for such an exploration to arise. Together they engender an examination into how what we know and who we are emerge and become our manners of teaching mathematics.

Feeling more settled in this new conceptual space of mine, I wondered what I should do next. Not having a clear direction in which to move my thinking, I looked to Stigler and Hiebert's question again but this time, how I might use it to prompt an exploration of my eco-centric one. It was in the midst of this back and forthing that I realized both questions spoke of concern for this place called, the mathematics classroom. It made sense for me then, that the place to begin was to begin with, place.

Sense of place complex? We tend not to think so, mainly because our attachments to places, like the ease with which we usually sustain them, are unthinkingly taken for granted. As normally experienced, sense of place quite simply is, as natural and straightforward as our fondness for certain colors and culinary tastes, and the thought that it might be complicated, or even very interesting, seldom crosses our minds...

~ Keith Basso[8]

Beginning with Place

Basso's description of the ways in which we create and sustain our relationships with places brings forth just how strong our connections to place are. In doing so, he reminds the reader how taken for granted, forgotten, unnoticed, or ignored the actual contextures and patterns that make a place a place really are. Primary, basic, and essential, sense of place is undeniably and always a critical part of every mathematics classroom.

If we perceive the world as a place in which we live as systems within systems, then the mathematics classroom, as place, is not constituted simply by the presence of four walls, some furniture, a teacher, students, and mathematics. Necessarily, this place includes the relations that evolve from the intermingling of teacher, children, their surroundings, and mathematics. If we think about how we come to know places, then our sense of a mathematical place in the classroom emerges from the spaces in which we perceive mathematics to arise and the forms it takes on. Put another way, the classroom as a mathematical place and how we connect with it not only comes from what we know and feel, but the kind of place it becomes grows out of the interactions we have with it. Thus, we do not define places; they do not define us. Rather, in dynamic interplay, we come to form together.[9]

As a teacher, I have always been committed to a holistic way of thinking about mathematics in the classroom, one that facilitated my growth in teaching mathematics and children's learning of it. But it was not until I had read Basso's book, *Wisdom Sits in Places* that I began to think ecologically about the mathematics classroom as place. Moreover, as ecological ways of being are not an everyday practice in our society or its educational systems, it is understandable that one would not think of the mathematics classroom in such a manner, much less be able to imagine what an ecologically coherent mathematical place might mean. Sense of place and ideas associated with place do not come about naturally or consciously for us. As a result, they remain hidden or invisibly embedded within our taken for granted manners of teaching mathematics.

Through my experiences in gaining a deeper understanding for how my teaching shapes the mathematics classroom and trying to create a sense of place that embodies ecological notions such as recursion, co-emergence, and fluid integration, I have learned that this kind of work cannot be achieved by what we think may be breaking out of conventional approaches. To do so in the attempt to get rid of all that undermines an ecological sense of place in the mathematics classroom would be naive and superficial. Systemic, ecological changes need to begin by first considering what place means to the mathematics classroom. It involves looking deeply and engaging in the complex, recursive process of identifying and questioning one's taken for granted conventions of thinking about and teaching mathematics – inquiring into and learning how these ways of being contribute to the sense of place that exists in the mathematics classroom.

Why is this so important? Does it really matter? I think so. Let me explain by describing the different senses of place for mathematics that I have come to know as both a learner and a teacher. These vignettes chronicle my growing understanding of place. They provide revealing glimpses into how deeply mechanistic, commoditized, linear, and disconnected my common sense of place for the mathematics classroom was and the challenges I faced in making it a cohesive whole.

Reflections Revealing the Fragility of Classroom Mathematics

It [mathematics] requires silence and neat rows and ramrod postures that imitate its exactitudes. It requires neither joy nor sadness, but a mood of detached inevitability: *anyone* could be here in my place and things would proceed identically.

~ David Jardine[10]

A Student's Place

Through my years of elementary schooling, I grew to believe that the spaces where mathematics existed in the classroom, the forms it assumed, and my relationship with it were clearly marked out by my teachers and had little if anything to do with my peers or myself. I came to know that at school, it is the teacher who makes mathematics happen. Just like TV programmers, I thought to myself, teachers ensured that mathematics began, ended, or reran at exactly the same time each day. They were conscientious not to let mathematics spill into any other programs of study such as science, socials, art, or language. And the only time when mathematics did extend past its designated slot was after school — if we had not completed our exercises during class.

Our lessons were similar to that of learning to catch a ball. First we watched the teacher demonstrate how to do the mathematics and then ready or not, the teacher threw problems up onto the chalkboard or to us in the form of a textbook. Scrambling to catch the mathematics, we madly recorded the mess of numerals and symbols in their correct linear fashion, practised, practised, practised, and then hopefully, tossed the mathematics correctly back to our teacher. On other days, we awaited the moments when timed drills, pop quizzes, and tests became the stage where we performed our proficiency and ability to juggle addition, subtraction, multiplication, and division facts.

I also learned that the teacher liked it best when mathematics happened not with other classmates but rather, silently in our heads, and figuring out solutions should not involve fingers, drawings, or counting. Never questioning but always reproducing, this is how my teachers and we students busily created and maintained a place of anonymity for mathematics in the classroom.

Encountering familiar issues in a strange setting is like returning on a second circuit of a Möbius strip and coming to the experience from the opposite side. Seen from a contrasting point of view or seen suddenly through the eyes of an outsider, one's own familiar patterns can become accessible to choice and criticism. With yet another return, what seemed radically different is revealed as part of a common space.

~ Mary Catherine Bateson[11]

Back on the Strip: My University Years

Towards the completion of my undergraduate degree in education at the University of Victoria, I was required to identify a specific teaching area that would become my focus for the remainder of my program of study. Based on my interests, it was a toss up between visual arts and mathematics education. I found it difficult to choose one over the other and so I chose the area that offered me boundless possibilities for complex and creative exploration. The two years of mathematics education and mathematics courses that followed made for what I considered to be two more journeys around the Möbius strip. The first trip, which I have already described, was my childhood experiences learning mathematics in school. The second, from the perspective of a teacher-to-be, and the third, from a learner of mathematics again is what I describe for you now.

During each of these returns, I found myself questioning and shifting my conceptions of what it meant to teach and learn mathematics. However, it was only afterwards that I realized it was on these journeys that I was visiting and revisiting the notion of mathematics' place in the classroom.

From the Side of Pedagogy

Now enrolled in mathematics education and approaching it not from that of a mathematics student but coming at it from the opposite side — one of pedagogy — provided me with the contrasting point of view necessary to reveal taken for granted conceptions I held about classroom mathematics.

It began in my first mathematics education class. In this course where the focus was on teaching mathematics, the professor engaged us teachers-to-be through modeling possible ways to develop children's conceptual understanding, actually experiencing hands-on minds-on activities for ourselves, and assessing student understanding through video analysis. In doing so, this professor dispelled many of my taken for granted assumptions about school mathematics; ones that included it as being an activity of simply doing tasks and solving problems quickly in one's head and that mathematics is best learned in isolation.

What became apparent to me was that as a teacher I needed to be creative in thinking about my manner of teaching as well as thinking creatively about mathematics itself. Forms of teaching which communicated to students that there are many ways to solve a problem, avoiding what was referred to as heavy-handed teaching that implied teaching-by-telling, and developing open-ended activities such as simple games or riddles that draw children into the complexity of the mathematics instead of repelling them from it became important foci in my growth as a teacher. Moreover, I realized that just as mathematics should be brought into being through the teacher, the students, and a variety of settings in the classroom, it was also important for teachers to enable learners to develop meaningful connections between the mathematics they study at school and the mathematics that occurs in their daily activities at home and in their community.

Learning from the opposite side of the Möbius strip — from the perspective of a teacher-to-be, I began to understand the impact that teachers have on the kinds of mathematics that arise in the classroom. And it was here that I began considering how I might enact a pedagogy that embodied a sense of connectedness for mathematics in the classroom.

Third Time Around

The following September, I found myself taking yet another trip around the Möbius strip. This time, again, from the perspective of a learner. Here in two of my mathematics courses I experienced first hand how it felt when mathematics took place in the dynamic ways that the first professor had described. These two professors made it clear to me that I would neither be given nor expected to memorize formulas or procedures. Feeling panicked, my first reaction was that I had registered for the wrong mathematics classes — how on earth was I to play the game if these professors did not show me which mathematics I was to toss back and forth?

Fortunately I persevered, continued to attend the classes, and for the first time in all of my years of learning, I was convinced that real mathematics was not a game of catch but rather something that is created and creative. Suddenly, for me, mathematics no longer took place in an anonymous world but with the world of human and natural contexts. We spent our time examining, questioning, and watching mathematical patterns emerge in different areas such as biology, economics, and everyday life. Learning in this manner provoked and enabled me to explore, devise, and create self-generated methods and mathematical formulations in place.

Common Space

By encountering familiar issues as M. C. Bateson describes and coming at them from the opposite side(s), we experience them as different or new. Then recursively, upon examination and questioning what we think to be distinct events, like the Möbius strip, we come to realize that the apparently disparate issues do not exist on separate planes but rather, exist within a common space.

For me, as a result of moving along what I perceived to be completely different planes was that my experiences of elementary mathematics learning, mathematics education classes, and university mathematics courses existed within a common conceptual space. Whether my experiences were that of a learner or teacher-to-be, they were all situated within the realm of sense of place for mathematics.

As a beginning teacher, I did not enter the classroom with a fixed image in my mind of mathematics as an activity that consisted of teacher demonstrations and student reproductions but instead, an image of mathematics as an ongoing engagement in which children and their teacher adventure in a world of knowing together.

SPACE WANTED
Looking to share a space with ecology. Interested in what ecologically coherent forms of teaching and learning mathematics could mean for the classroom. Can move in IMMEDIATELY.

HELP WANTED
"How can we break out of our conventional approaches and imagine more productive alternatives?"[1] Reply to mailbox: T1I9M9M7S

VACANCY
Seeking one teacher to teach grades 2/3. Separate room. Shared facilities.

(continued from page 36)

3 FACES OF ECOLOGY

M. C. Bateson[2] identifies three "faces" or realms of ecology: empirical, enviromental, and systemic. The author defines empirical ecology as biological, meteorological, and geographical studies that focus on understanding how the planet is changing and how these changes affect the interrelationships of the world's natural systems. The environmental face of ecology is concerned with identifying the level of impact that our ways of living have on the earth's systems. This area of ecology also involves the development of solutions for environmental problems that minimize harmful stress on the earth. It is within the systemic realm of ecology where mathematics teaching and learning can be most radically explored. This is because systemic thinking focuses on seeking "the pattern which connects"[3] a system or systems together as interacting wholes.

A "system" could be a teacher or a student. It could also be a collective group such as a mathematics class, a school, and so on.

The connecting pattern or patterns that interrelate one system to another encompass the forms of knowledge, actions, and identities that result from the ongoing interactions within the systems as well as the ways in which these dynamics are sustained by the systems themselves.

Focusing on relational qualities makes it possible to view the world as an integrated whole; a dynamic and fluid unity in which all living and social-cultural systems are interconnected. The world is not conceptualized as a collection of separate entities but instead, as a highly complex unity in which all systems are interrelated and therefore, interdependent.

It seems sensible that when looking from an ecological perspective, mathematics teaching could be conceived as similar to that of students' mathematical learning.[4] Mathematics teaching then, as a fluid and complex process, implies that it exists always in relation to the ongoing interactions of the students, the mathematics, and the material and nonmaterial environment of the classroom.

In light of this, it is from a systemic perspective that the following question is posed:

> ***In what ways can systemic manners of thinking about mathematics education enable forms of teaching and spaces for learning mathematics to possess an ecological sensibility?***

Many, and perhaps *most* teachers begin their careers with the conviction that they will avoid those teaching practices that they found unhelpful or inappropriate when they were students. .... However, ... most beginning teachers quickly find themselves settling into patterns of teaching that are strikingly similar to the ones they intended to avoid.

~ Brent Davis, Dennis Sumara & Rebecca Luce-Kapler[5]

Traditions, Tensions, and Transitions

Turning The Soil: The Beginning Years of Teaching

I began teaching second and third grade children in Richmond, British Columbia. When I entered the teaching field, I distinctly remember being eager to inspire a more connected sense of place for mathematics in the classroom and at the same time, being careful not to become another Mrs. Fibonacci. Mrs. Fibonacci, a storybook character, is an elementary school teacher who loves and lives mathematics to such an extreme that she makes learning an unbearable nightmare for one particular child in her class by turning everything into a mathematical problem that needed to be solved. For the main character, learning mathematics becomes a curse he cannot escape:

> "What if this keeps up for a whole year? How many minutes of math madness would that be?"
> "What's your problem" says my sister.
> "365 days x 24 hours x 60 minutes," I snarl.[6]

Like all beginning teachers, I devoted enormous amounts of time to preparing my lessons. I read journals for mathematics teachers, went to workshops searching for new ideas, collected real world materials to connect the children's mathematics with familiar contexts, and designed interactive mathematical tasks that would engage every child in my class; all the while, being careful not to cast any curses. But even though the children, their parents, and my colleagues seemed to be pleased with my efforts, I did not feel as if I was accomplishing what I had set out to do. My teaching and the children's mathematical learning still seemed disconnected.

What we conserve, what we wish to conserve in our living, is what determines what can and what cannot change in our lives.[8]

Insight, I believe, refers to that depth of understanding that comes by setting experiences, yours and mine, familiar and exotic, new and old, side by side, Learning by letting them speak to one another.[7]

In the months that followed, I searched and scrutinized my mathematics program to find the source of my unease. I poured over the curriculum guides to be certain that I was teaching the correct concepts and the skills for the different grade levels. I revised the order and adjusted instructional sequences so that they moved more efficiently. I continued to tweak or elaborate the content of my lessons, depending on the needs of my students. Looking at the program as a whole, I felt that I was engaging the children in mathematical work that enabled their learning to be both hands-on and minds-on and that I was making space for mathematics to be integrated with other subject areas. Unable to find any obvious problems, I continued to proceed along the current course.

Then, several months later, I started to question the kinds of relationships that existed between my teaching and the children's learning of mathematics. I took a reflective step back and examined my mathematics program for a pattern or patterns that connected the children's mathematical learning spaces together as a whole. In doing so, taken for granted ways of teaching began to emerge. I discovered that these were not only rituals unique to myself, but surprisingly, they were matter of fact ways of being for my colleagues too... even those of my schoolteachers. For me, these teaching practices had simply become THE *way to facilitate children's learning in the mathematics classroom. Intrigued and wanting to take a closer look, I wrote these manners of being down on paper. And as I did this, it became apparent to me just how matter of fact they were and how deeply embedded in my teaching these shared facilities*[9] *had become.*

- BEFORE BEGINNING ANY LESSON, SORT THE CHILDREN ACCORDING TO THEIR GRADE LEVEL. THEN, TEACH EACH GROUP A DIFFERENT MATHEMATICS LESSON.

- WHEN PLANNING THE CURRICULUM FOR THE SCHOOL YEAR, SIMPLY DIVIDE THE MATHEMATICAL CONCEPTS AND SKILLS FOR EACH GRADE INTO TEN EQUAL PARTS. NOW, ALLOCATE ONE OF THE TEN SCHOOL MONTHS TO TEACHING ADDITION, ONE MONTH TO SUBTRACTION, ANOTHER MONTH TO MULTIPLICATION, AND SO ON, UNTIL THE END OF JUNE.

- ALWAYS MAKE SURE THAT CONSISTENT AMOUNTS OF TIME ARE GIVEN TO MATHEMATICS LESSONS. SCHEDULE IT IN REGULARLY EACH DAY (E.G., EVERYDAY BETWEEN RECESS BREAK AND LUNCH HOUR).

- MATHEMATICAL CONCEPTS SHOULD BE TAUGHT SEQUENTIALLY; FROM AN INFORMAL, CONCRETE STAGE TO MORE FORMAL, ABSTRACT ONES. TEACHING SHOULD FACILITATE THE STUDENT'S (THE AUTONOMOUS CHILD) CONSTRUCTION OF KNOWLEDGE IN A CONCEPTUAL TO PROCEDURAL TO RELATIONAL ORDER.

After reading this "must do" list, it was apparent that the sensibility of wholeness and flow that I desired for the mathematics classroom did not exist. Instead, was one that possessed rigid, mechanistic, and disconnected qualities. These could be seen, enacted in my scheduling of lessons at same time every day, my taking inventory of the mathematics curricula and then packaging them up into discrete units of instruction, teaching separate grade-specific lessons, and my always doing so in a manner that proceeded from the concrete to the abstract. The kind of place that I had intended to root and the one that had actually become embedded were in contradiction to each other. What served as tried and true rituals for teaching mathematics had unthinkingly become that which was furthering of what Berry refers to as the cultivation of discrete parts without respect or responsibility for the whole.[10] My teaching actions not only dismembered mathematics for the children but on another level, I had also dismembered mathematics from itself. I say this because one might argue that my efforts to teach for the students' conceptual then procedural then relational knowledge could be viewed as in keeping with facilitating connected understandings. However, despite the fact that I did this in my teaching within each of the concepts and procedures, I was still teaching the concepts as separate parts and attention was not paid to enabling the students' connections among concepts, procedures or mathematical topics. One might also argue that put together these individual units of instruction came to form a complete mathematics program. This might be true; however, the parts still were not fluid or dynamic but rather, discrete and static.

In taking this reflective step back, I saw that it was not enough for me to design a well, put together mathematics program and I wondered what I might do to engender a clear sense of flow in the mathematics classroom. Even though I could see how some of my invisible or assumed ways of teaching were undermining this, I did not know what kinds of re-rooting (conceptual or otherwise) were necessary.

What I learned however, was just as Basso[11] describes, place is not something that can be taken for granted — not even in the mathematics classroom. Place is primary and basic yet at the same time, it is far more complex than had originally crossed my mind. If places are indeed created and sustained through interaction, then the mathematics classroom as place exists only in its being. And, it could be said that what distinguishes one mathematics classroom from another is its sense of place. Further still, it is then the kinds of mathematics that emerge from one's teaching and from children's learning that define the actual terrain of a mathematics classroom.

EDUCATION

"JUST HAND THEM DOWN THE MATHEMATICS" ... OR NOT?!

WHEN MATHEMATICS is imagined and enacted as objectified, static knowledge that is to be traditionally passed down from one generation to the next, the teaching and learning of mathematics is disabled from ever becoming anything else. Under the air of hand-me-downs, it is easy to understand why mathematics is taught and learned out of a sense of obligation or contempt rather than a sense of open desire or wonder. It is also understandable why mathematics is all too often considered as that which is to be mastered rather than that which is to be understood. In commoditizing mathematics, we make absurd, the possibility for us as teachers and our students to perceive mathematics as anything else but a fixed entity. In this way of conceiving mathematics, we make it inconceivable for school mathematics to become something other than just a collection of hand-me-downs.

The embeddedness of these images within one's taken for granted ways of thinking about mathematics not only makes it natural for us to assume mathematics to be an inanimate thing, but in doing so, locates mathematics as that which exists — out there. Given this mindset, it is not surprising that a teacher would feel impelled to set the class on a straight and narrow, one-way course to become collectors of mathematics. Given this mindset, it makes sense to ingrain the ritualistic practice of acquiring mathematics into school mathematics unit and lesson plans, methods of assessment, and enacted in the classroom; product oriented practices that focus on desired, expected, or even measurable outcomes of instruction — that after instruction — the students will have mastered the mathematics taught in the lesson before moving on to the next part of the curriculum. Of course, the ways in which students are instructed to take possession of their mathematical hand-me-downs of concepts, skills, and even attitudes may vary. Still, teaching by telling, engaging students in hunting for, having them seek out hidden mathematics in real world contexts, and even explorations de-

(continued on page 24)

(continued from page 8)
signed for students' discovery of mathematics all keep alive, the tradition of handing down mathematics.

Moreover, when product-oriented ways of thinking about school mathematics are coupled with a back to basics mentality, the teaching and learning of mathematics becomes subjected to the weigh scale of how much in regard to the amount of mathematical facts and skills that students are to learn and little or no emphasis is placed on such things as their mathematical thinking or conceptual understanding. Given this mindset, mathematical processes such as those identified by the National Council of Teachers of Mathematics[12] as: problem solving, reasoning, communicating, connecting, and representing, would likely be deemed by most teachers as not essential. Viewed as additional knowledge,[13] teaching that attends to the development of mathematical processses would then depend on whether students have acquired first, the prespecified mathematical facts and skills with which to process the mathematics.

The point here is that when students are taught to learn mathematics in the tradition of hand-me-downs and as a product oriented matter of collecting, hunting down, or retrieving pieces of knowledge, this creates the impossibility for mathematics to be taught and learned in ways that enable it to arise as living and animate.

Now, identifying the limitations of how mathematics exists in the classroom and the possibility of it becoming something else is all fine and good. But in doing so, means that the conversation does not end here. Rather, it opens up a whole host of questions that require further consideration such as:

- How might an ecological way of thinking help in reconsidering such taken for granted perceptions and re-imagining a more responsive view for the teaching and learning of classroom mathematics?
- What shifts in thinking are necessary to conceive classroom mathematics as something other than a line of hand-me-downs from teacher to student?
- What would it mean if we assumed mathematics to be embodied?
- How might problem solving, reasoning, communicating, connecting, and expressing be understood as integral to the learning of mathematics and not as additional knowledge?

Settling In(to AnOTHER Space)

I met Jennifer at the coffeehouse. She greeted me as I pulled up a chair to join her. Reaching for the newspaper that was lying on the table, she leafed through it and then stopped when she got to the **Letters to the Editor** section. Jennifer began reading one reader's response to an earlier article, **JUST HAND THEM DOWN THE MATHEMATICS ... OR NOT?!**

MOVING THINKING SPACES AND ASSESSING OLD FURNITURE

IN RESPONSE TO LAST WEEK'S EDITORIAL: I TOTALLY agree with the arguments that were raised. And the questions for sure, are important ones in making positive changes in the math classroom. My worry, though, is that real changes can't happen if this job of rethinking and re-imagining classroom mathematics is approached with the attitude of getting rid of or simply adding onto what's already there. What the editorial didn't mention was that it's not about taking ecological ways of thinking and coordinating them like new pieces of furniture into a rundown living room with the hope of updating one's mind-space by making it look more current. Really, a major part of moving from invisible and mechanistic places of knowing to ecological ones is about rediscovering the all too familiar furnishings that have been set about (classroom) mathematics and asking, how well do these furnishings go with the space?

All for new spaces,
Joel

"... getting rid of... adding onto what already exists... no. Definitely not," said Jennifer as she read the letter.

And so she continued, reading bits of it silently in her head and every so often, sputtering out particular words or phrases.

"Exactly," said Jennifer as she looked back up at me. "Enabling deep changes in teaching mathematics isn't about

> changing out of certain approaches and slipping into new ones."

She agreed with Joel that what she needed to do was to examine her mathematics teaching from where she (conceptually) stood now. Jennifer wondered what she might see and see differently from an ecological perspective. What kinds of furnishings had become so comfortable and integral to her mathematics teaching that they were now permanent and perhaps, invisible? Jennifer questioned whether they would even suit an ecological mind-space. And more to the point, she was anxious to know what kind of place, what kind of *oikos*[14] she was engendering for her students' mathematics. But where might she begin?

For several days, Jennifer thought about the specific vantage point from which to position her thinking and to examine these issues. What kept emerging for her was a distinct and recurrent theme. The theme was recursive in the sense that it was the place where Jennifer found her thinking returning again and again. Moreover, it was not until this moment that she recognized the ever presence of this location. Here was the place where she existed both in and away from the classroom. When she described it to me, I immediately named it her in-between space. Not because it was a space of indecision for her but rather, the in-betweeness had to do with how her teaching and her research co-emerged and co-evolved. For Jennifer, teaching and research neither existed as separate entities nor did they move sequentially from one to the other as she once thought.

> "As a pre-service and a beginning teacher I understood research to be something that was conducted by others who were not part of the classroom and its function was to produce theory which in turn, became a tool that I could use in my practice of teaching mathematics," she said. "But now, what comes to my mind is an image of teaching and researching as continuous interaction and dynamic growth... whereby teaching and researching flow into and give rise to each other because of each other. So teaching, exists as praxis and not simply the establishment and maintenance of one's teaching practice."

As I listened to Jennifer describe how her view of mathematics teaching and research had changed, I realized that to characterize her in-between space as the location where the two met or intertwined would be to miss the meaning altogether. They did not meet. They were each other. For Jennifer, teaching mathematics and her study of it were inseparably connected. In a complex yet circular manner, Jennifer considered them to be interacting, co-evolving systems; that is, each system as a necessary part of the other. Furthermore, the distinction she made regarding her shift from teaching as practice to teaching as praxis revealed that mathematics teaching as praxis was not simply a routine that one performs but that which requires one's active engagement; teaching as praxis then, implies a way of being that is critically reflective and reflexively responsive. Evoked from within an ecological realm is the importance of being ever mindful of how our knowing, actions, and identities in teaching mathematics arise recursively from that which we have already lived and therefore, become embodied in how we are living now, and how we will live. In other words, teaching as praxis recognizes the simultaneity and the complex circularity of that which occurs in one's teaching is also necessarily enfolded in all that interacts with it. This space certainly was not an in-between space. It was not a conceptual space located somewhere in the middle of teaching and research. Really, it was an*other* space.

> "I see this kind of reflexivity as necessary in attempting to understand how it is that my teaching and research give rise to each other," said Jennifer.
>
> "So, if you had to describe in your own words, the guts of it all, how might you do so?" I asked.
>
> "Simplifying the complex? Hmm... let's see... I suppose I would have to say that for me, teaching learners mathematics and learning what it means to teach mathematics flow together."

And it was in this spirit and in this other space that Jennifer began the process of researching her teaching by encircling it within ecology.

Jennifer naively assumed that by placing her teaching within the realm of ecology, immediate answers to her questions would be revealed. However, as the days passed, it was only her impatience that became apparent. Frustrated, she picked up a book that she had been reading and turned the page. There in black and white was the reminder she needed.

> **Letting myself get written by a place.** Bodily scars as the agelines in the droopy skin on the backs of my hands betray. Legibilities of having, once again, lived-through.
> **Sitting squat. Spending time. Waiting. Reserve. Quiet. Composure. Patience. Letting the boredom arrive. Wasting time. Doing nothing with great deliberateness. Collecting dry bones. Boredom: this is one great little demon we have banished from the discourse of authorship and expression and self-annunciation. Deliberately spending time in the old place, feeling through moist weaknesses:[15]**
>
> > Perception of opportunities requires a sensitivity given through one's own wounds. Here, weakness provides the kind of hermetic, secret perception critical for adaptation to situations. **The weak place serves to open us to what is in the air.** We feel through our pores which way the wind blows. We turn with the wind: trimmers. An opportunity requires... a sense... which reveals the daimon of a situation. The daimon of a place in antiquity supposedly revealed what the place was good for, its special qualities and dangers. The daimon was thought to be a *familiaris* of the place. **To know a situation, one needs to sense what lurks in it.** (Hillman, 1987, p. 161)[16]

Although Jardine's description details how he readies himself to write, this practice of dwelling and keeping watch was exactly what Jennifer needed to do. It was obvious to her now that she did not know what aspects of her mathematics teaching needed closer examination; so to search for a deeper understanding when she could not articulate what exactly it was that she was seeking a deeper understanding for became a ridiculous endeavor. Jennifer decided it was best if she wandered back to that other place where teaching and learning and what it means to teach mathematics flowed together. There she waited patiently, dwelling, and all the while, keeping watch for what lurked in it.

Surfacing & Noticing

So Jennifer, what came out?

I wouldn't say that anything really came out... it was more like bubbles making their way to the surface of the water and then bursting in the moment of recognition. You know, like, there's one... and another one! Once you've caught sight of one of them, all of a sudden, there they are! And you see them for the first time — but not because they magically appeared —they'd been there the whole time and were only invisible because you hadn't been able to notice them before.

Do you mean that there were qualities about your teaching that you were unaware of?

Well, yes. Even though my focus was on ecology and the mathematics classroom, up until this point, it wasn't specifically about my teaching.

Does it need to be?

I think at some point it has to be. You see, it was only when I began questioning the kind of place I was creating for my students' learning of mathematics that I recognized the need to move deeper — when I say deeper, I mean delving into my teaching — not just being attentive to what's emerging in my teaching but also to what's already developed and become the roots of my mathematics teaching. Had I not realized this, any change that I might make would most likely be superficial because I hadn't considered the whole of my teaching — I hadn't dwelt long enough to notice what was there. So like Joel mentioned, my ecological ways of being would be at best, add-ons.

Okay, just a minute. You figured out that you needed to look at these aspects of your teaching but you still hadn't figured out what you needed to be noticing, right?

Exactly. More dwelling. What I did know was that by exploring my teaching in these ways, I might be able to articulate why I was so uneasy about the sense of place I was creating with my students in the mathematics classroom... why it didn't feel right. But still I had no idea of what specifically I in my teaching I should examine.

So what did you do?

Well, this time, I took the whole of my mathematics teaching back to where I'd been — you know, into the space of language and languaging.

Then what?

As I immersed my thinking into theoretical literature on language and the pervasive nature of language, I began by asking myself, so how do these concepts inform my understanding of teaching mathematics?

And... ?

Certain key ideas began to pool together. In fact, some were direct quotations from the books and articles I'd been reading.

Such as?

Well, like:

Language THINKS US as we think with and in language.
- Martin Heidegger[17]

"METAPHOR is not a mere embellishment; it is the basic means by which abstract thought is made possible."
- GEORGE LAKOFF AND RAFAEL NÚÑEZ[18]

The map IS NOT the territory.
~ Alfred Korzybski[19]

"...[language] does much more than construct concepts about the events and things in the world; it TELLS us what sorts of concepts we ought to construct."
~ Neil Postman[20]

And these ideas pooled together because they were all...

They all had to do with the metaphorical nature of how we think.

So it was just these specific authors' works?

Yes and no. It was these four ideas that kept making their way into my mind-space of language and languaging but these weren't the only ideas I was thinking with. They'd emerged from other authors' work such as Sfard, Abram, Capra, M. C. Bateson, van Manen, Jardine, Orr, Maturana, Varela, Lakoff and Johnson.... but yes, it was these specific meta-voices that really helped to pinpoint my position of noticing — so that I could begin to examine my mathematics teaching.

And so what was your position or perspective of noticing? Can you explain it to me?

Let me see. Well for starters, it was directed towards the way in which metaphors become embodied in our forms of knowing, our actions, and our identities. Remember when we were talking about language?

Yes.

By metaphorical language I mean, the spoken, written, spatial — how phenomena are portrayed to exist, temporal — how time is conceptualized, and symbolic forms of communication that distinctively structure one's teaching. So briefly, there it is. My point of noticing was to examine the metaphorical furnishings, if you will, of teaching mathematics and directly related to this, the kind of place that I was bringing forth in the classroom.

Do you think that metaphorical forms of communication really impact one's mathematics teaching in such a profound manner?

Yes I do. Just take a moment to think about it: Theoretical FOUNDATIONS, *instructional* UNITS, *conceptual* FRAMEWORKS, *learning* JIGSAWS, *cognitive* STRUCTURES, SCAFFOLDING, BUILDING *knowledge, and so on. Metaphors. We are constantly reading them, hearing them, using them, and thinking with them. So usual they come to be just as Joel wrote — permanent fixtures in one's mind and over time, we no longer notice their presence. Unquestioned, these metaphors become embedded in our taken for granted language; language used to conceptualize mathematics teaching and learning — language that directly impacted my teaching, my students' learning, and the kind of mathematics that emerged.*

Wait a minute. Go back to what you just said... are you saying that metaphors were affecting the kind of mathematics that emerged? Surely, math is math. Metaphors have nothing to do with the nature of mathematics.

Oh, but they do. My previous mind-space predetermined the mathematics content that was to be taught and learned. Because of this, I wasn't aware of or didn't pay any attention to emergent kinds of mathematics[21], integrative mathematics[22], or individual-within-collective mathematics.[23]

So what?

Well, it does matter. If you're not aware of how metaphorical language is thinking you as you think with and in it, it's difficult to understand how it affects our thoughts and how it becomes an enabling and disabling feature in our ways of being.

Okay. So make sense of it for me at the classroom level. I still want to know what surfaced for you when you interrogated your teaching. Let's go back to my original question, what came out? What surfaced as a result of your dwelling in this theoretical realm and in this other space of yours?

Well, what came to the surface of my consciousness were metaphors that carried with them very vivid meanings of how I thought, taught, and identified my role in the mathematics classroom. For the first time, I began to understand how unconscious these metaphors were.

What do you mean?

Well, because I hadn't realized how pervasive and taken for granted these were, the metaphors existed below the surface of my consciousness. They were definitely there but up until this point, I didn't have a theoretical way to examine my teaching and so, it was impossible for me to notice these metaphors. They were simply matter-of-fact ways of conceptualizing and enacting my teaching.

For example, how you mechanistically separated and organized the mathematics content for each grade level into teachable units of instruction?

Yes.

And so it was these embedded metaphors that you began to notice rising to the surface?

Exactly. And how it happened was just as I described for you at the start of our conversation — like bubbles making their way to the surface of the water and punctuated by them bursting as soon as I spotted them. What I also learned as I searched to identify my taken for granted metaphors was that they arose from the theoretical languages of my undergraduate mathematics education and Ministry documents, teacher texts, and mathematics literature I was working with. You see, it was here that the relationship between my activities in reading and writing and how I envisioned my work in the classroom... that is, planning, teaching, and assessing children's understanding of mathematics became clear. Almost instantly, the metaphors that'd been totally invisible were now so obvious. Because they were visible, I could see the metaphorical images embedded in everything from the way I imagined the mathematics class to my conceptualization of mathematics curricula.

What kinds of metaphors? Give me some examples.

HANDS-ON MINDS-ON

CHILD-CENTRED approach

The **AUTONOMOUS** child

Knowledge is **CONSTRUCTED**

A GUIDE

The teacher's role is to **FACILITATE** learning

The **TEACHER** is the **INITIATOR**

Children are unique **INDIVIDUALS**

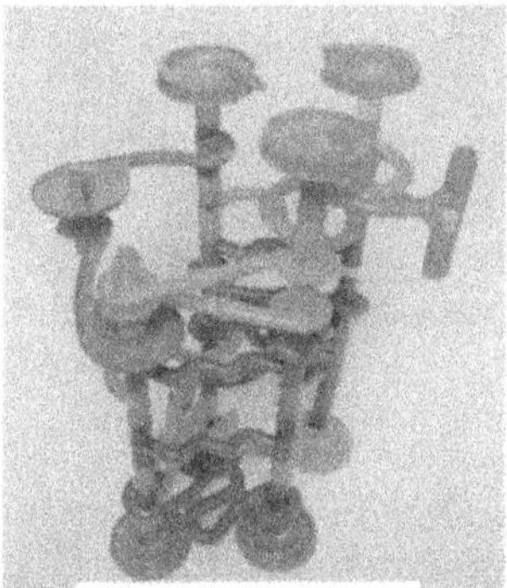

MENTAL SCHEMAS

REALITY IS CONSTRUCTED by the individual

EXTERNAL environment

COMMUNICATING

PROBLEM SOLVING

REASONING

MATHEMATICAL POWER

REPRESENTATIONS

CONNECTIONS

INSTRUMENTAL understanding

RELATIONAL understanding

FORMAL understanding

PROCEDURAL understanding

CONCEPTUAL understanding

SCAFFOLDING

from **FROM CONCRETE TO ABSTRACT**

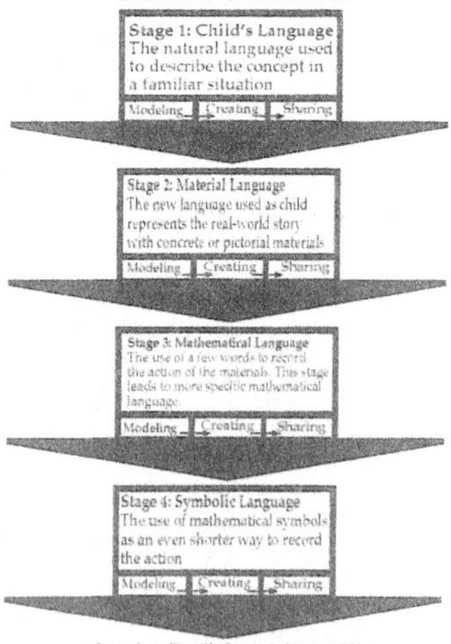

(based on Reuille Irons & Irons, 1989)

Okay. For instance, take the **MATHEMATICS CLASS**. *As I imagined it, the class was composed of the students and myself as* **AUTONOMOUS INDIVIDUALS**. *Anything outside the individual was considered to be part of the* **EXTERNAL** *environment.*

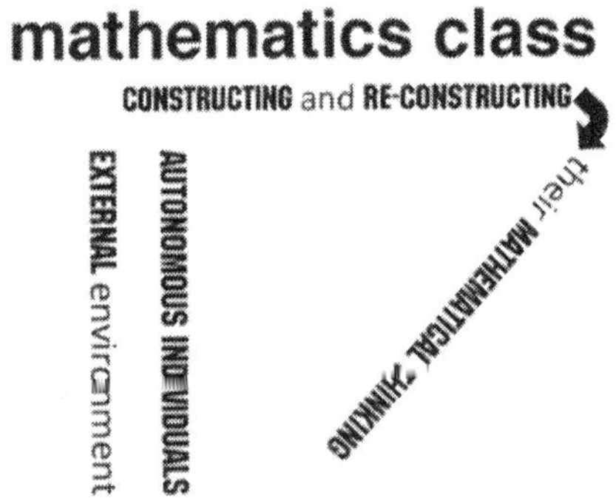

MATHEMATICS *itself existed as a* **CONNECTED YET FIXED** *body of knowledge. It was made up of* **SEPARATE STRANDS** *of algebra, geometry, numbers and operations, measurement and so on.*

I envisioned **MATHEMATICAL UNDERSTANDING** *to develop in a* **UNI-DIRECTIONAL AND HIERARCHICAL MANNER**; *beginning at an informal,* **CONCRETE** *stage and then moving towards* **HIGHER** *levels of more formal,* **ABSTRACT** *stages of understanding.*

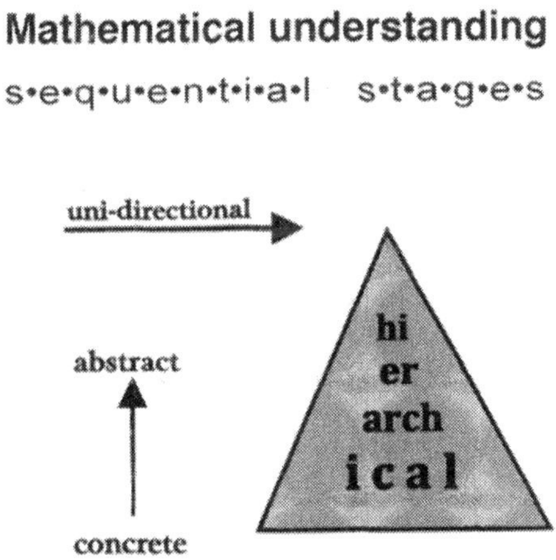

mental KNOWLEDGE STRUCTURES

the BIGGER the BETTER

mathematical understanding

MATHEMATICAL UNDERSTANDINGS *were knowledge* **STRUCTURES** *or* **FRAMEWORKS**. *For me, the bigger and more elaborately constructed the structure, the better the* **INDIVIDUAL'S** *understanding.*

Based on the premise that if knowledge was about building structures, **TEACHING** *mathematics for me became an activity of directing my students' thinking towards* **PRE-DETERMINED** *learning outcomes in regards to what they should know and facilitating the ways in which they should* **CONSTRUCT** *such knowledge.*

teaching mathematics

MOVING STUDENTS ON A COURSE towards PREDETERMINED learning OUTCOMES →

curricula

jigsaw puzzles

mathematical skills

mathematical concepts

And as well, **CURRICULA** *were* **JIGSAW PUZZLES** *for teachers to assemble by piecing together concepts and skills set out by the BC Ministry and other* **STANDARD** *mathematics documents.*

I agree that these are very strong metaphors... mechanistic, linear, and hierarchical ones to be sure. But did they really affect the way you taught mathematics? And if so, how?

Having made them visible, I asked myself this same question: Were these metaphors simply figures of speech or were they more than that? Were these metaphors truly powerful forms of language? Language that was not only shaping how I perceived mathematics teaching and learning to be but also, profoundly impacting how such events came to be. So I turned my attention — a little apprehensively, I must admit — towards examining if and how, these metaphors I had identified existed in my forms of teaching.

And…?!

Well, rather abruptly I came face to face with the notion that all knowing really is doing and all doing really is being. What's more, how unthinkingly natural it all is.

You see, because I viewed mathematics to be a connected yet fixed body of knowledge and curricula were puzzles to be assembled, my goal in creating an integrated mathematics program was to connect the different pieces of mathematics together to produce a logical and coherent picture for the students. In thinking so, it made sense for me to insert their lessons in — in a piecemeal fashion for an hour each day between recess and lunch. And in keeping with the view that mathematical learning was sequential and hierarchical, because I taught a multi-age class, it made it necessary to sort the students according to their grade level and teach two different lessons. My role as their teacher was to guide each student's learning in a manner that enabled the student to construct sturdy frameworks of understanding; ones that began with concrete foundations of experiential knowings upon which more formal, symbolic representations were built. I even remember being asked on several occasions as to how I defined myself as teacher.

And, how did you describe yourself?

I was the children's facilitator... the initiator of learning opportunities or to stay with the metaphors, the provider of building materials.

After all of this, what was your reaction in realizing that your teaching was indeed enactions of the mechanistic, linear, and hierarchical metaphors that you were thinking with?

If you'd asked me this question before finding this all out, I would've said that I'd probably be shocked, disappointed... even horrified if I discovered that my teaching contradicted the ecological perspective that I thought I was embracing.

What do you mean? You weren't shocked, disappointed, or horrified? Really. I'd think that burying one's head in the sand, so to speak, would be a common reaction.

Strangely enough, it was more of an affirmation... finally being able to recognize the metaphors that had existed for so long beneath the surface of my consciousness and then to see the embodiment of them in my teaching — relief through affirmation... yes, that's what it was. For some time, I'd had a hunch, a gut feeling, that the connected sense of place I was trying to create in the math classroom didn't exist... but I couldn't put my finger on it as to why.

But you had, had you not? You had identified rituals in your teaching that were linear, mechanistic, and hierarchical?

Sure, I was able to point to teaching actions that I'd unthinkingly inherited and see them as problematic... such as planning a program by dividing and ordering the mathematical concepts and skills for each grade into a September through June sequence. But doing so only indicated forms of teaching I deemed as undesirable. It still didn't provide me with any kind of understanding as to what was giving rise to them or how I might go about creating a more ecological sense of place for my teaching and my students' learning.

Yes, that's right.

It was only when I moved deeper into my teaching and examined my metaphors... Hmm.... how can I describe the process to you.... For me, this process was very much like, fingering the contaminated wound[24] *— explicating my metaphors and then watching them fester — how the metaphors were being enacted in my mathematics teaching. Yes, that's an accurate image.*

How horrible that sounds!

Well, yes. The image impresses my experience as incredibly uncomfortable — even painful. And in some ways, it was. Finding out that you're doing exactly the opposite of what you wish to do is definitely distressing and uncomfortable but at the same time and in a different way, I by no means considered the study of teaching as pathologic.

Although that's how I think most people would interpret your description.

I know… but no. To come to this place in my thinking was critical. The uncomfortableness of it all wasn't a prompt for identifying and remedying a problem so much as it was an opening for me to arrive at a new place of understanding. You see, I considered the dis-ease of these events to be integral and vital to my growth in teaching mathematics. It was because of dwelling in this mind-space that I was able to grab hold of what I could before only express as a hunch or a gut feeling and now, I was able to put words to it and finally say, here it is!

For sure, making the invisible visible. When you describe your metaphors and explain how they gave rise to your forms of teaching, it clarifies the point you've been trying to make; that the metaphors and metaphorical patterns with which a person thinks has everything to do with that person's teaching of mathematics and the sense of place that's created in the classroom. What also becomes clear is that even though you wanted to create a connected sense of place through creating an integrated mathematics program, the metaphors you unconsciously rooted in your mind critically disabled the possibility for an organic or ecological kind of integrative mathematics to come about. The metaphors only allowed for mechanical piecemeal forms of teaching and learning — definitely not those that are dynamic, flowing, or unpredictably open. It's exactly as you expressed earlier in our last conversation — that teaching learners mathematics and learning what it means to teach mathematics really do flow together. That all said and going back to Joel's question, now that you had figured out what furniture didn't suit your ecological living space, how did you go about finding furnishings that might?

Deeper Yet

With her enactions of deeply rooted metaphors now exposed, Jennifer realized that not only were her conceptions and her mathematics teaching necessary parts of each other, they were in fact, each other. It was impossible for her to consider metaphors as merely figures of theoretical speech. She now understood the invisible yet ever-present manner in which these metaphors specified her patterns of thinking and her teaching of mathematics. In a very real way, the metaphors and metaphorical manners were Jennifer's rituals for place-making in the classroom and awkwardly, she knew that her taken for granted ways of teaching mathematics did not engender the ecological forms she wished to enact.

Settled into this other conceptual space, Jennifer's new understandings brought about another realization: that in the moments of making sense of the constraints of her metaphors and knowing that she wanted to enact ecologically coherent ones, she could not simply change by exchanging what she was thinking or doing in the classroom for something else.

AGAIN, I am reminded that learning what it means to teach mathematics is not an automatic process. It is not smooth, it is not straightforward, and it certainly does not appear on demand.

KEEPING WATCH WHILE DWELLING REQUIRES PATIENCE.

And before she could re-imagine metaphors that were ecologically sound and work towards rooting them into her classroom praxis, Jennifer knew that she needed to exercise a mindful kind of patience in order to move even deeper into this other space so that she could question, assess, and provoke shifts in her thinking. It was here in the act of dwelling that Jennifer found herself wondering about conceptions of mathematics, mathematical understanding, and the different ways in which they are brought into being.

How is mathematics conceived? And in doing so, what kinds of being does it become?

PROCLAIMED THE *QUEEN OF SCIENCES* PRISTINE IN ITS SACRED CUSTOMS OF PRECISENESS AND LINEARITY... MATHEMATICS IS ABSTRACTLY ELOQUENT AND REFINED... IT OFFERS US TRUTHS SO ABSOLUTE SO PURE SO UNAMBIGUOUS... IN ITS PREDICTABILITY AND CERTAINTY AND EXPLICIT INFALLIBILITY. IT IS NEITHER TO BE QUESTIONED NOR TO BE HELD ACCOUNTABLE FOR ANYTHING EXCEPT ITSELF. TREASURED HEIRLOOMS OF LOGIC AND RATIONAL KNOWLEDGE... UNIVERSAL AND TRANSCENDENT... MATHEMATICS EXISTS IN THE REALM OF OBJECTIVITY... IT IS NEUTRAL AND LIVES WITHOUT REGARD FOR US... OUR BELIEFS... OUR VALUES... OUR ACTIONS... OUR CULTURAL WAYS... IT LIVES OUT THERE AS THE FOUNDATIONS OF THE UNIVERSE

ORDERING AND STRUCTURING THE UNIVERSE

•

•

•

the
flowers
the snowflakes
the ferns and the trees

creative and beautiful

the stars and the planets
the shells of snails
the orbits
us

•

•

•

rooted in our ways of being

not **OUT THERE** *but not only* **IN HERE** *either... mathematics arises from our biological, social, cultural human being in the world*

a science of patterns... mathematics is not **a THING** *but a way perceiving... making visible, the otherwise invisible* [1]...

a living, growing system openly structured of nonlinear dimensions and never an **END PRODUCT**

a SHAPE-SHIFTER *of number, of form, of action... never static but dynamic, mathematics is fluid and ever-changing... quantitative or qualitative... or imaginary... material or mental... purposeful or fanciful... complex and diverse... interacting, shaping, and being shaped...*

multiversal and constitutive... mathematics exists in the realm of (objectivity)... not inert and not neutral...

mathematics lives in relation with us... with-in our beliefs... our values... our actions... our cultural ways...

mathematics lives here... in dynamic relation with our human being.

Does the way one portrays mathematical understanding matter? And if so, how does it shape one's perception of what it means for learners to understand mathematics?

Theoretical Portraits of Mathematical Understanding

Within mathematics education, there exists an array of models and theoretical points of view that address aspects of mathematical understanding from the very general to the very specific. Two themes that are inherent within and across these three collections of literature are that of cohesion and tension. While there is general agreement among mathematics educators of what "good" mathematical understanding entails, the ways in which educators portray the nature of mathematical understanding, how it comes to be or should be developed, and the forms that arise create contrasting tensions amongst one another.

First, several works from perspectives situated within what is considered to be part of a constructivist realm are showcased. Here, one gets a sense of what it means to frame mathematical thinking and learning within this theoretical discourse and how doing so portrays understanding to be the building and rebuilding of mental schemas. The second collection of research involves theoretical viewpoints that move away from linear or constructivist minded frameworks and towards more holistic and dynamic interpretations of learners' mathematical understanding. Finally, works that are located within an enactive realm and strive to illuminate mathematical understanding as co-emergently complex phenomena are examined and discussed.

EXAMINING MATHEMATICAL UNDERSTANDING

FROM A CONSTRUCTIVIST PERSPECTIVE

A MODEL OF INTELLIGENCE AND FORMS OF MATHEMATICAL UNDERSTANDING

Skemp's (1979) model of intelligence offers a qualitative means for describing individuals' mathematical understanding. The two-level, cybernetic model (see Figure 1) consists of two internal systems: delta-one and delta-two. Delta-one is a sensori-motor system that directs an individual's physical mathematical actions based on information received from the external environment. Delta-two serves as the site where construction and reconstruction of an individual's mental mathematical schemas take place. It is this process of schema construction and reconstruction that allows for the mathematical functioning of delta-one to occur. Thus, it is "the construction and testing by delta-two *within* delta-one of the schemas and plans that delta-one must have to do its job." (Skemp, 1979, p. 44) It is here in delta-two where Skemp identifies mathematical understanding as developing. The specific ways in which these internal systems function together is evidenced through one's *instrumental, relational,* and *formal* or, *logical* forms of mathematical understanding.

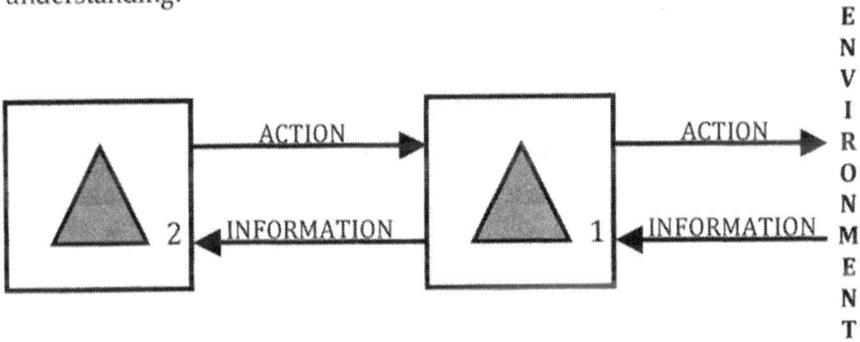

Figure 1. Skemp's two-level cybernetic model of intelligence (based on Skemp, 1979).

Instrumental Understanding

Skemp explains instrumental understanding as being a function of delta-one. This form of mathematical understanding enables a person to correctly apply previously learned procedures to solve mathematical problems. Instrumental understanding is in essence, the learning of what to do with the mathematics. It does not however, enable the learner to develop conceptual grasps for interpreting why a method works or what the symbols might mean (Skemp, 1978, 1979). For example, by remembering the words and the order of the letters in the acronym, BODMAS (**B**rackets, **O**f, **D**ivision, **M**ultiplication, **A**ddition, **S**ubtraction), a person can carry out the correct sequence of numerical operations for solving complicated calculations.

The way in which learners are able to develop instrumental mathematical understanding is through rote methods of demonstration and further practising of a particular procedure or set of skills until it becomes routine. Although the cognitive structures in delta-two that result from instrumental learning enable a person to manipulate mathematical symbols and rules, the person's actions remain restricted because the connections that are formed in the delta-two schemas exist only as relationships between symbols and rules, not among mathematical concepts. The extent to which a person is able to apply his or her instrumental understanding to different mathematical contexts then remains limited to combining and performing procedures in the prescribed sequence that these were learned.

Relational Understanding

Relational understanding is evidenced by a person's appropriate strategies for solving mathematical problems. This form of understanding involves the individual making sense of why particular methods of mathematics may work and why others may not be effective when solving certain problems (Skemp, 1978). In other words, relational understanding implies the learner's knowing of what to do and, why certain mathematical actions prove to be effective. Skemp describes the manipulation of mathematical concepts and schemas as a function of delta-one while the individual's conscious or unconscious reflection of these concepts and schemas takes place in delta-two.

Unlike instrumental understanding, relational understanding gives rise to schemas that connect mathematical concepts with procedures. This form of learning develops as an individual alternates between activities of interacting mathematically in the external environment and mentally reflecting on these experiences. As relational understanding can only be achieved through the individual's conceptual integration of mathematics, this process requires more time than does instrumental learning through rote methods. However once acquired, relational mathematical understanding is seen to be more flexible because such knowledge is connected to mathematical concepts and not to specific contexts, it can continue to develop. And unlike instrumental learning where an individual recognizes a mathematical problem and then applies and performs a prescribed procedure to solve for it, an individual with relational understanding can derive mathematical procedures through conceptualizing or comprehending the task at hand. Mathematical symbols do not exist as abstract objects on which an individual performs actions but rather, these carry meaning for the individual in that the symbols are objects to which conceptual understanding is attached and enable the construction of connected schemas of concepts and skills.

Formal or Logical Understanding

Skemp characterizes this form of understanding to be present when an individual consciously connects symbolic mathematical language together with meaningful ideas and logical reasoning. This occurs as either a delta-one activity within both delta-one and delta-two. If a person possesses delta-one logical understanding, the learner is able to reflect on his or her mathematical actions through an if/then type of rationalization; that is, "*if* I perform the correct methods to solve a given problem *then* the result should be correct" type of thinking. On the other hand, Skemp describes logical understanding that takes place in both deltas as being when an individual shows through formal mathematical demonstration or proof that the applied mathematics makes sense through inferences that connect the given premises of the problems to established mathematical axioms or theorems. This type of formal functioning that occurs within delta-two enables the individual to become aware of the connections between delta-one and delta-two activities and establishes consistency between the individual's mathematical schemas and solutions.

Last, within each of these three forms of understanding — instrumental, relational, and logical-formal, there can be *intuitive* and *reflective* dimensions to the learner's mental functioning. Intuitive mathematical functions are characterized as spontaneous processes that occur in delta-one and do not necessarily include the delta-two system. When intuition occurs in both deltas, this gives rise to the unconscious reflection of the individual. However, it is only when the individual is consciously aware of his or her activities in both the first and second deltas that this process is considered to be reflective.

MATHEMATICAL UNDERSTANDING AS CONNECTING AND CONSTRUCTING NETWORKS

Hiebert and Wearne (1992, 1996) apply a constructivist definition founded in cognitive psychology (Brownell, 1935; R. Davis, 1984; Hiebert & Carpenter, 1992; Lesh, Post, & Behr, 1987) to define their view of what they consider mathematical understanding to be. They refer to it as the learner's development of mental connections and formation of networks that serve as representations of mathematical ideas. For example, Hiebert and Wearne would consider a well-connected understanding of multi-digit addition to be a network that consists of the child's knowledge of concepts regarding place value, basic facts, and the ability to generate effective procedures to deal with the task at hand. They believe the process by which understanding of mathematical ideas occurs is an unpredictable, recurrent, and nonlinear progression. Furthermore, the flexibility of an individual's mathematical understanding is seen as an indication that the learner has constructed mental networks that have many points for external information to enter and to trigger the individual's successful adaptation, acquisition, and retrieval regarding appropriate strategies to solve mathematical problems.

Linking Written Symbols with Understandings

Site 1: Interpretation and Development of Meaning for Symbols

- What do these symbols mean?
- What am I being asked to do or find?

This site focuses on written symbols and the ideas or objects that they represent:

- **numerals** as quantities (e.g. 5 km or 5 apples)
- **operations** (i.e., addition, subtraction, multiplication, and division) as actions on quantities in the natural world
- **signs** as relationships between or among quantities (e.g., =, <, and >)

Site 2: Developing Meaning for Rules

Establishing what to do and why

This site includes the use of manipulatives when introducing rules or procedures as an important step in illustrating how a rule works and connects the symbolic answer to the concrete solution.

Site 3: Producing an Answer

This site focuses on:

- making an estimation
- taking action (i.e., applying the chosen procedure)
- examining the solution based on previous estimation
- applying or relating the symbolic problem back into an informal context (e.g., Would your answer hold true when put back into a real-world context?)

Figure 2. Summary of Hiebert's three sites for linking written symbols with understanding.

Building Bridges to Connect Informal and Symbolic Mathematics with Student Understanding

Hiebert's independent and collaborative research seeks to understand the relationships that exist between children's conceptual understanding of mathematical concepts and their external abilities to recall and modify existing procedures, construct suitable methods, adopt prescribed rules, and use symbolic mathematical language with understanding (Hiebert & Wearne, 1993). A commonality throughout Hiebert's research concerns itself with previous studies (Carpenter, Hiebert, & Moser, 1983; Lindquist, Carpenter, Silver, & Matthews, 1983; Hiebert & Wearne, 1986) and similar arguments raised by other mathematics educators such as Usiskin (1996), Pimm (1987), and Carraher, Carraher, & Schliemann (1987) regarding students' lack of connection between symbolic mathematics found in the classroom and that which occurs in their everyday life. Hiebert's work emphasizes the need for less formal mathematical representations to serve as a means by which meaningful connections for children's understanding and application of symbolic mathematics are developed.

As well, the author asserts that in order for students to use the symbolic language of mathematics to their advantage, essential connections regarding their informal, experiential knowledge must be recognized by teachers and made explicit to their students. Hiebert's framework (1989) (see Figure 2), which identifies three critical sites for linking written symbols with understanding, highlights the necessity for children to develop meanings for the ways in which symbolic mathematics can be used as a powerful language in solving problems. This model makes clear the need for students to make the symbols work for them instead of working with the symbols. Stressing the importance for the learner's utilization and integration of out-of-school mathematical behaviours[1] with school mathematics, Hiebert's model clearly identifies that the final stage of mathematical understanding should not be the learner's ability to perform symbolic mathematics but rather, the child's meaningful understanding for and their ability to reintegrate their use of symbolic mathematics into a variety of settings.

Problemetizing Children's Learning of Mathematics

Teaching methods considered to foster rich, connected schemas of mathematical understanding are ones that enable children to *problemetize* their mathematics. Hiebert et al. (1997) distinguish problemetizing mathematics as being different from problem solving approaches to learning that imply teacher demonstration and children's imitation of identifying key words in a problem, selecting an appropriate method, and performing prescribed calculations to solve a task. Rather, problemetizing of mathematics is viewed as facilitating deep mathematical understanding because it focuses on students making sense of and developing meaningful relationships between their mathematical ideas and mathematical actions. Hiebert et al. explain this approach to learning mathematics as elucidating:

> ... reflective inquiry as the key to integrating ideas and actions. Problematic situations, and methods of inquiry used to resolve them, elicit ideas *and* actions. This is what distinguishes problemetizing from traditional problem solving in which an acquired procedure is applied. (1997, p. 24)

Within these types of lessons, the teacher strives to structure learning opportunities that are not only interesting to the students but also introduce important mathematics to them. Students are expected to make sense of the mathematics and methods they employ through discussions led by the teacher that interrogate the effectiveness of particular methods, as well as exploring different ways of representing their understandings

[1] For example, interpreting, judging, devising, estimating, and evaluating.

through written, verbal, objects, pictorial, symbolic, and informal means of mathematical language. As students seek to resolve problematic situations such as determining the difference between 72 and 39, the teacher's facilitating them into actively generating, adopting, or reflecting on mathematical strategies and ideas allows the students' learning to be "tasks, and discussions... [that] connect with where students are and that are likely to leave an important mathematical residue." (Hiebert et al., 1996, p. 17)

THE CONSTRUCTION OF MATHEMATICAL UNDERSTANDING

R. Davis (1984) explains mathematical understanding in a manner similar to Minsky and Papert's (1972) view; that mathematical understanding is present when an individual is able to integrate a new idea into a larger structure of previously constructed ideas. R. Davis (1992) uses the metaphor of assembling a jigsaw puzzle to illustrate his view:

> ... that one assembles ideas in one's mind much as one assembles a jigsaw puzzle. Each new candidate piece, like each new idea, can be used only if it fits into the aggregate of pieces that have previously been assembled. (p. 228)

R. Davis states that if we consider mathematical proofs or even Skemp's reflective, logical understanding to be the results of mathematical activity, then mathematical understanding must be taken to be the result of children's working with mathematics. In this way, R. Davis (1992) considers mathematics to be a result of children's understandings:

> Instead of starting with *mathematical* ideas, and then *applying* them, [teachers] should start with *problems* or tasks, and as a result of working on these problems the children would be left with a residue of *mathematics*... that mathematics is what you have left over after you have worked on problems. (p. 237)

The Teaching of Mathematics

R. Davis poses a similar argument to one found in Hiebert et al.'s work (1996, 1997) that stresses that rather than teaching children mathematics through methods of showing and telling, connected understanding only develops when children have established for themselves a reason for doing mathematics. Solving tasks in this manner provides opportunities for students to decide whether they will employ already established methods or construct mathematical procedures on their own. This is explained below:

> Instead of telling students what to do, and leaving them wondering about why one does it this way, the new approach helps students

> understand the task or the goal, and gives students the responsibility for inventing ways to solve the problem. (R. Davis, 1992, p. 238)

In examining the role of the mathematics teacher, R. Davis and Vinner (1986) claim that if we believe students build up their mathematical schemas through constructing and reconstructing ideas based on their previous experiences, then mathematics teachers play an integral role in students' learning. On the other hand, the teacher's instructional actions cannot be viewed as those that ultimately determine the ways in which students form their mathematical schemas. Mathematical schemas are assumed to be constructed from and always influenced by the child's previous mathematical experiences.

R. Davis and Vinner raise another issue with respect to student errors in mathematics. They assert that student errors should not necessarily be considered an indicator of lack of understanding but could be, in fact, the student's retrieval or selection of an inappropriate mathematical idea. Since teachers cannot determine what mathematics a child will or will not choose to retrieve, this becomes a responsibility of students to be aware of their mathematical understanding. R. Davis and Vinner encourage learning settings that engage students in nonroutine mathematical problems (Schoenfeld, 1985; Silver, 1994) as a way for learners to further construct their mathematical understanding and learn skills in monitoring their mathematical actions.

MATHEMATICAL AMBIGUITIES

R. Davis and Vinner identify five sources within students' school and out-of-school experiences that may obscure students' conceptual understanding in mathematics. These are as follows: the language of mathematics, assembling mathematical representations from pre-mathematical fragments, building mathematical concepts, the impact of specific examples, and children's misinterpretation of mathematical experiences. They argue that teachers should not try to exclude ideas from contexts outside of mathematics because these, as all other mental representations, serve as necessary parts in children's assembly of pre-mathematical ideas (Dubinsky & Lewin, 1986). So in a similar manner to Sierpinska's argument[2] for understanding the importance of epistemological obstacles and Dubinsky and his collaborators' use of genetic decomposition as a method for mathematics instruction,[3] R. Davis and Vinner advise that we should not attempt to prevent children from developing mathematical misconceptions but rather, enable them to be aware of misconceptions in their thinking and how overcoming

---

[2] As discussed on p. 95-99. See also, Sierpinska (1990, 1994).

[3] As discussed on p. 101-105. See also, Cottrill et al. (1996).

these is necessary in making sense of mathematics.

Mathematics as a Language

In North America, the English language is viewed as the linguistic base with which students enter school and from which they begin to build mathematical ideas. The English language is also identified as a source of many difficulties in terms of students' mathematical understanding. Pimm (1987) explores the possible reasons for this confusion.

Pimm (1987) puts forth the notion that just as English has specific ways in which it functions as a language, mathematics too possesses its own linguistic register and has "a set of meanings that is appropriate to a particular function of language, together with the words and structures that express these meanings" (p. 75). He explains that within the mathematics register there are *specialist terms* or, words that hold specific meanings in the context or discipline in which they are functioning. Durkin and Shire (1991) refer to these specialist terms as lexical ambiguities of mathematical language and make further distinctions between these words by classifying them into four subcategories: *homonymy, polysemy, homophony,* and *shifts in applications*. Homonymy describes words that have the same form as in English but imply different meanings in mathematics. For example, the word, "leaves" does not signify "leaves" on a tree or the verb "to leave", but rather, describes the subtractive action in mathematics. Polysemy characterizes mathematical words that may have two or more different but related meanings to their English definitions. For instance, the word, "product" in English, can be defined as "something that has been made", and in mathematics, takes on a similar meaning, "a quantity obtained by multiplication". Homophony is defined as two or more distinct words that have identical pronunciation but entirely different meanings — as observed in the words, *two, too,* and *to* or, *sum* and *some* or, *pi* and *pie*. Shifts in applications are similar to what Pimm (1987) describes as *notational metaphors* and these are mathematical symbols, that in combination with other symbols, convey particular meanings. Here we can see that the number, 5 can be applied in mathematics to communicate the nominal meaning of "the number five", the ordinal meaning of "the fifth number", the cardinal meaning of "1, 2, 3, 4, 5", or the visual representation of "5".

An example illustrating this difference between English and mathematics linguistic registers is observed in the use of the word, any. In ordinary everyday contexts, this word is most often taken to mean 'some' yet in mathematics, this word implies 'every' such as, "is *any* odd number prime?" Interpreting this question in a nonmathematical manner, one's answer would most certainly be yes, as seen in the case of the number five or seven. However, comprehending the word as meaning the latter, one would have to answer no, as not all odd numbers such as nine, have only factors of one and itself. A second example is located in our use of number words that function in mathematics not only as adjectives as in *one* house, but can also exist as nouns such as when we speak of prime *numbers*, implying that numbers have distinct qualities. Moreover, given the mathematical fact that, *four fours* are *sixteen*, number words also operate as adjectives and nouns.

Metaphors and the Construction of Mathematical Concepts

Metaphorical usage of English words in mathematics is evident in elementary school when children learn to "carry" when regrouping numbers in addition, to "borrow" when renaming numbers in subtraction, or making reference to the "face" when identifying surfaces of 3-D objects. Pimm (1985) identifies three types of metaphors existing in mathematics: "structural", "idiosyncratic", and "standard" or "conventional" metaphors. Structural metaphors refer to the ways in which symbols are arranged together and take on different meanings. Idiosyncratic metaphors are metaphors invented by the user to make sense of a mathematical concept or idea. Standard or conventional metaphors are used and understood by many people. Examples include "having" for positive numbers, "owing" for negative numbers, a mathematical function is a "machine", or an equation is a "balance". These metaphors serve as tools with which students think and build images about mathematical concepts. In other words, they are "functioning images... which [can] connect the ideas of mathematics with objects and processes that [students] feel they know and understand" (Pimm, 1987, p.97). Pimm cautions teachers that while metaphors are valuable tools in helping children conceptualize mathematical ideas, it is necessary for teachers to help students to define the usefulness of a metaphor by exploring it in many different contexts. By doing so students develop an understanding of where and when their use of metaphors is appropriate and when it may be a mathematical act of over generalization.

In the same sense this does not mean that given linguistic ambiguities in mathematics, teachers should try to teach for all possible meanings or misconceptions that may arise when students learn a particular mathematical concept. Rather, it is to be expected that at different stages of a child's learning, some aspects of mathematical concepts will be fully developed while other aspects may be partially explored or not at all.

Besides metaphorical over generalization, there are other times when children take specific instances and construct generalizations from them. This phenomenon that R. Davis and Vinner (1986) see as impacting on students' mathematical understanding, is observed when students learn about the multiplication of whole numbers and form the conclusion that multiplication of any values always produces a greater product. While the authors argue that it is fine for teachers to let learners maintain their partial understanding of a concept as long as the contexts in which the students are applying it is appropriate, R. Davis and Vinner also stress the need for teachers to provide settings that provoke children to engage in reconstructing their understanding of a concept. In the example of multiplication, reconstruction of the concept would become necessary when the children begin to work with decimals, fractions, and negative numbers. Finally, R. Davis and Vinner explain that because of the ambiguities that exist in mathematics, students may misinterpret what mathematics is being taught and thus, teachers need to be cognizant that children's focus on unnecessary or extraneous aspects of a given concept can also lead to mathematical misconceptions.

HERSCOVICS AND BERGERON'S ANALYTIC FRAMEWORK

In their effort towards enabling teachers to teach for children's mathematical understanding, Herscovics and Bergeron (1981, 1982) developed their analytical framework (see Figure 3), to describe key characteristics of particular mathematical concepts such as *number* (Herscovics & Bergeron, 1988b; Herscovics, Bergeron, & Bergeron, 1986a), *length and surface area* (Héraud, 1988), and algebraic concepts such as *slope* (Dionne & Boukhssimi, 1988). Herscovics and Bergeron (1988a) assert that the development of individuals' conceptual understanding should always begin in their physical, concrete world. They advocate for teacher practices to be those that value not only children's written answers but place an equal emphasis on children's thinking processes. Herscovics and Bergeron consider their framework to be a tool that can aid in epistemological analysis of mathematical concepts. Moreover, by accounting for the different components of Herscovics and Bergeron's model when planning instruction for a particular mathematical concept, teachers can design and provide richer mathematical learning settings.

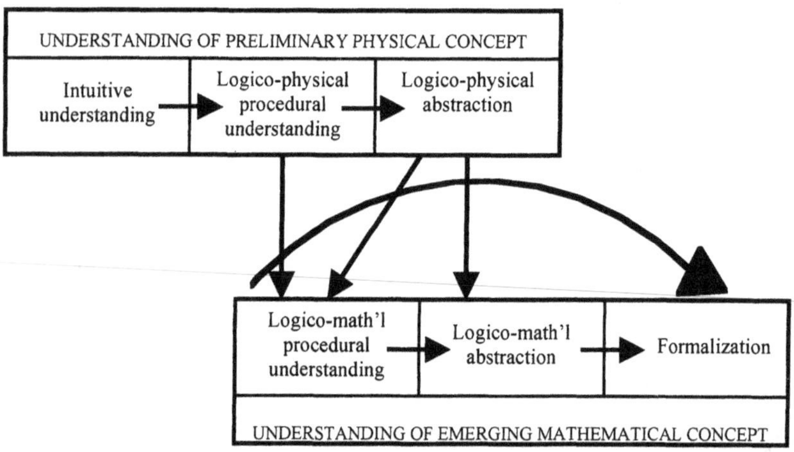

Figure 3. Analytic framework of mathematical processes (based on Herscovics, 1989).

The model is divided into two partially sequential but non-hierarchical tiers. Herscovics and Bergeron's two-tiered model conceptualizes mathematical understanding as being a partially sequential process that begins first with the individual's intuitive understanding of physical concepts and then develops through a series of levels into an abstract, mathematical concept. The arrows within the model indicate the forms of logico-physical and logico-mathematical abstraction that are generated from the individual's preliminary physical concepts. The other arrows show that an individual's understanding of a mathematical concept does not require all three parts within the first tier. This assertion is supported by Herscovics and Bergeron's observations of young kindergarten children who are seen to have mastered counting procedures and the formalization of the concept of number but have not yet comprehended all of the invariances regarding quantity and rank.

The first tier is identified as, "understanding of preliminary physical concepts" and consists of three distinct components of understanding: *intuitive, logico-physical procedures,* and *logico-physical abstraction.* Here, intuitive understanding is based on the individual's visual perception that provides non-numerical approximations. Logico-physical procedural understanding is evidenced by an individual's ability to relate his or her intuitive knowledge through 'physically acting out' mathematical concepts. The authors describe logico-physical abstraction to be when an individual synthesizes and constructs meaningful relationships such as reversibility or generalizations between physical mathematical concepts (Herscovics & Bergeron, 1983, 1984).

The second tier encompasses another three components: *logico-mathematical procedural understanding, logico-mathematical abstraction,* and *formalization* that Herscovics and Bergeron consider to be integral parts of comprehending mathematical concepts. Herscovics and Bergeron define procedural understanding to be when a learner relates preliminary physical concepts that underpin logico-mathematical procedures, such as counting methods for determining quantity or rank by using them appropriately in a given context (Bergeron, Herscovics, Bergeron, 1986; Herscovics et. al, 1986a). Logico-mathematical abstraction refers to the individual's construction of connecting logico-mathematical invariants together with related logico-physical invariants to form generalizations, such as coming to know that the commutativity of addition as a property applies to all pairs of natural numbers (e.g., 4+3 and 3+4 both equal 7) (Herscovics, Bergeron, & Bergeron, 1986b). Finally, Herscovics and Bergeron characterize formalization as an individual's activity of axiomatizing and producing mathematical proofs. At an elementary level, children's discovery of axioms and finding logical mathematical justifications would be taken as indicative of formalization. They also consider formalization to include the enclosing of a mathematical notion into a formal definition as well as the use of mathematical symbolization for such notations. For example, this type of formalization of procedural understanding is observed when a child writes out a sequence of digits while counting.

MATHEMATICAL UNDERSTANDING AS TAXONOMY

Mathematics educators Pegg and Currie (1998) agree with Piagetian views that assume older children learn in a qualitatively better way than do younger children because they have more developed mental structures. At the same time however, Pegg and Currie take a different stance with respect to observing and analyzing students' mathematical understanding. They support the view put forth by Biggs and Collis (1982) as well as other researchers (Blake, 1978; Hallam, 1967) that different methods rather than ones that generalize students' academic performances based on Piagetian cognitive developmental stages are necessary in order to provide detailed descriptions regarding students' learning within discipline-specific contexts.

The prestructural, unistructural, multistructural, relational, and extended abstract levels in Biggs and Collis' SOLO (Structure of the Observed Learning Outcome) taxonomy are described as being "isomorphic to, but logically distinct from, the stages of preoperational, early concrete, middle concrete, concrete generalization, and formal operational, respectively" (Biggs & Collis, 1982, p. 31). There are also four dimensions within each of the five levels that are used to further categorize student responses. They are as follows: working memory capacity, operations relating task content with cue or question and response, and general, overall structure (see Figure 4). In keeping with Piagetian models that focus on hypothetical cognitive structures (HCS), SOLO also forms a concrete to abstract framework. In contrast, unlike HCS, that characterizes the individual in terms of age and stage of development, the SOLO taxonomy does not attempt to describe the learner, but rather, the quality of the learner's response(s) within a specific context and, in terms of the theory's levels and dimensions. By adapting elements from Biggs and Collis' theoretical framework of the SOLO taxonomy, Pegg and Currie assert that in this way, they are able to analyze students' mathematical understandings in a more detailed manner than allowed by Piaget's developmental stages.

Developmental base stage with minimal age	SOLO description	1 Capacity	2 Relating operation	3 Consistency and closure	4 *Response Structure* (*Cue* — *Response*)
Formal Operations (16+ years)	Extended Abstract	*Maximal:* cue + relevant data + interrelations + hypotheses	Deduction and induction. Can generalize to situations not experienced	Inconsistencies resolved. No felt need to give closed decisions—conclusions held open, or qualified to allow logically possible alternatives. (R_1, R_2, or R_3)	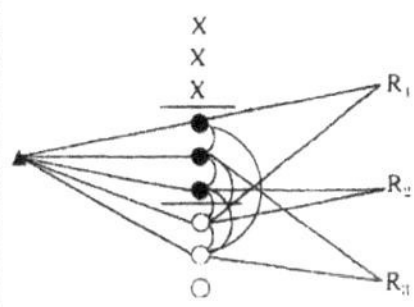
Concrete Generalization (13-15 years)	Relational	*High:* cue + relevant data + interrelations	Induction. Can generalize within given or experienced context using related aspects	No inconsistency within the given system, but since closure is unique so inconsistencies may occur when he goes outside the system	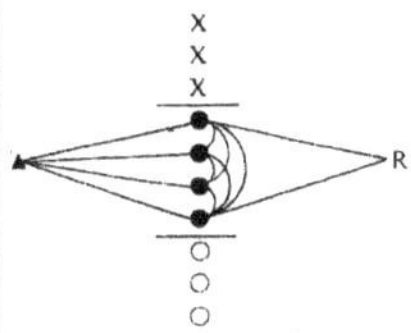
Middle Concrete (10-12 years)	Multistructural	*Medium:* cue + isolated relevant data	Can "generalize" only in terms of a few limited and independent aspects	Although has a feeling for consistency can be inconsistent because closes too soon on basis of isolated fixations on data, and so can come to different conclusions with same data	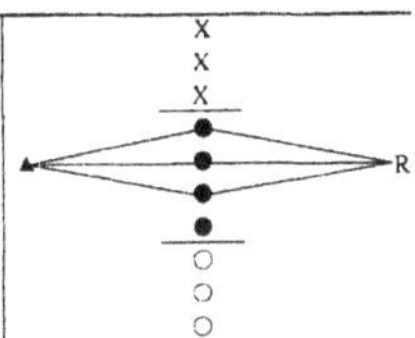
Early Concrete (7-9 years)	Unistructural	*Low:* cue + one relevant datum	Can "generalize" only in terms of one aspect	No felt need for consistency, thus closes too quickly: jumps to conclusions on one aspect, and so can be very inconsistent	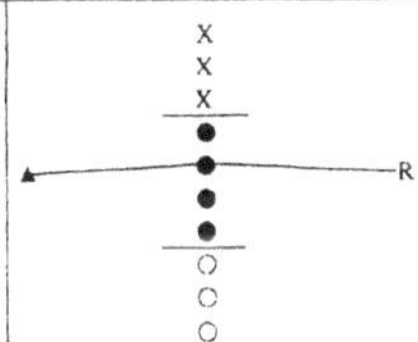
Pre-operational (4-6 years)	Prestructural	*Minimal:* cue and response confused	Denial, tautology, transduction. Bound to specifics	No felt need for consistency. Closes without even seeing the problem	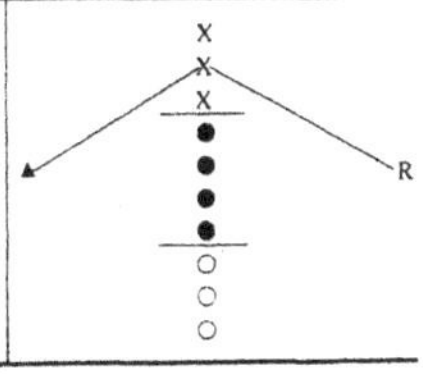

[a] Kinds of data used: X = irrelevant or inappropriate; ● = related and given in display; ○ = related and hypothetical, not given.

Figure 4. Biggs and Collis' (1982)[4] five levels and brief descriptions of each level of the SOLO Taxonomy and the taxonomy's loose correspondence with Piaget's stages of cognitive development.

[4] This diagram was published in *Evaluating the quality of learning: The SOLO taxonomy (structure of the observed learning outcome)*, p. 24-25, Copyright Elsevier (1982).

The Application of the SOLO Taxonomy to the Analysis of Students' Geometric Understanding

Just as Biggs and Collis created the SOLO taxonomy because they deemed Piaget's developmental stages as not appropriate for looking at children's understandings, mathematics educators Pegg and Currie integrate elements of SOLO to further elaborate on the van Hiele theory (van Hiele, 1986; van Hiele-Geldof, 1984) (see Figure 5) in order to analyze children's geometric understanding. Pegg and Currie's main criticism concerning the van Hiele model of geometric thought is that the model "cannot address questions posed outside of the direct notions of properties of figures, class inclusion, and deduction about which the theory is explicit" (1998, p. 334-335). What is common to the SOLO taxonomy and the van Hiele model is that they both derive from Piagetian roots. The difference between the two models is the integration of SOLO with the van Hiele model, geometric understanding shifts focus from whether or not the student has "mastered" levels of understanding (Pegg & Davey, 1998), to categorizing students' responses in a polychotomous manner whereby answers are grouped with similar characteristics and reflect various stages of cognitive growth.

Level 1:
Figures are identified according to their overall appearance. Properties play no explicit role in this identification process.

Level 2:
Figures are identified in terms of properties and are considered to be independent of one another.
***Level 2A:**
Figures are identified in terms of one single property, such as the length of sides of a figure.
***Level 2B:**
Several properties are identified but exist in isolation of one another.

Level 3:
Relationships between previously identified properties of a geometric figure are now established.
*The student is able to order the properties so that one or more properties give rise or imply other properties.

Level 4:
Deduction is understood and students can develop mathematical proofs.

Figure 5. Pegg's (1997) description of four levels of the van Hiele theory. (*) indicate SOLO-derived adaptations made to the van Hiele theory.[5]

[5] There is a confusing linguistic mismatch of the levels between the van Hiele model and Pegg's (1997) revision of it. The van Hiele model begins with the basic level of geometric understanding and continues on to levels one through four. Pegg's level 1 is assumed to be an elaboration to that of van Hiele's basic level. Pegg's level 2A and 2B are taken to be complementary to level one in the van Hiele model. As well, Pegg's level 3 and level 4 are understood as elaborations of the second and third levels of the van Hiele model.

Given the descriptions above, Pegg incorporates notions of response levels drawn from the SOLO taxonomy to the van Hiele theory and, by doing so, elaborates on the level descriptors to enable more inclusive criteria against which to compare and analyze students' geometric understanding. Specifically, Pegg (1998) and, Pegg and Davey focus on the elaboration of the second and third levels of the van Hiele model. Here, unistructural responses as well as multistructural responses are considered to be within the concrete symbolic mode and are associated with level 2 thinking. Biggs and Collis describe the concrete symbolic mode as being when individuals use and learn symbol systems. Typically, this level of functioning takes place in late primary through secondary school years and requires the individual to be able to internalize and generate representations of objects and events as words or images. Next to prestructural, Biggs and Collis (1991) describe unistructural responses as the second most concrete form of understandings. A unistructural response involves the learner only having to comprehend the given task or question by relating the question with a response that incorporates one of the concepts found within the problem. For example, given a picture of two equilateral triangles and asked to respond in terms of what is the same about the two figures, a unistructural response could be that both figures have three corners, three corners being one of many possible similarities.

The authors define multistructural responses as when an individual is able to focus on two or more relevant concepts at one time. While considered to be a more sophisticated level of understanding than that of unistructural responses, a second characteristic of multistructural responses is evidenced by the individual's comprehension of the concepts as being separate and not as related ideas. An example found within this level of understanding could be a student who identifies, given a picture of the two equilateral triangles, that both triangles have three corners each and that each of the triangles' three sides are of equal length. Level 3 responses are evidenced when students generate an overview of, or identify the important elements within a task in order to form an appropriate generalization. Here, relational responses of a concrete symbolic manner are deemed by Pegg and Currie as characteristic of level 3 thinking wherby, "[t]he relational response requires, in addition (to accessing a number of concepts), an overview of relevant concepts while being able to monitor the process or task from beginning to end, thus allowing for a logically complete conclusion" (Pegg & Currie, 1998, p. 337-338).

MATHEMATICAL UNDERSTANDING AS OVERCOMING EPISTEMOLOGICAL OBSTACLES

Understanding as a Process and an Act

Sierpinska (1990) examines what it means for mathematical understanding to be conceptualized as both a process and an act. She agrees that "[u]nderstanding is achieved *slowly*, along with the accumulation of properties of objects, examples and development of concepts concerning relations between classes of concepts" (cited in Lindsay & Norman, 1984, p. 438). In this manner, mathematical understanding is the process by which an individual's constant construction and reconstruction of ideas and meanings results in the establishment of connections between mathematical concepts. Sierpinska also draws on Ricoeur's (1989) notion regarding the dialectic nature of the process of understanding and as acts within a process. Sierpinska (1990) adds that although she agrees with Ricoeur's general idea of the dialectic between an individual's understanding and explaining as "starting with a guess and developing through consecutive validations and modification of the guess" (p. 26), it is difficult to directly apply this model to the comprehension of mathematical concepts. To do so necessitates the individual's experience in working through a variety of situations "because the understanding of a concept is not normally reached through reading a single text. It demands being involved in certain activities, problem situations, dialogues and discussions, and the interpretation of many different texts" (Sierpinska, 1990, p. 26). Sierpinska integrates the ideas of both Lindsay and Norman, and Ricoeur, to conceptualize mathematical understanding as a process and act of constant construction, generalization, and resynthesis of ideas and relationships between concepts through a spiraling "process" of dialectic interpretation.

> *Processes* of understanding are seen as lattices of *acts* of understanding linked by various reasonings (explanations, validations) and a (relatively) 'good' understanding of a given mathematical situation (concept, theory, problem) is said to be achieved if the process of understanding contained a certain number of especially significant acts, namely acts of overcoming obstacles specific to that mathematical situation. (Emphasis added, Sierpinska, 1994, xiv)

A Historico-empirical Approach

Sierpinska (1987) explains that in the act of understanding mathematics, new ways of knowing are established. Distinguishing her theoretical work as being different from other models of mathematical understanding that focus on levels of understanding (Herscovics & Bergeron, 1988b; Pirie & Kieren, 1989; van Hiele, 1986), cognitive structures (Dubinsky & Lewin, 1986; Lesh, Landau, & Hamilton, 1983), or the dialectic coupling of procedural and relational forms of understanding (Sfard, 2000; Skemp, 1978), Sierpinska classifies her research as being that of a historico-empirical approach. This approach examines students' understanding of mathematics from a perspective that focuses on the "obstacles to understanding encountered both in the history of the development of mathematics and in today's students." (Sierpinska, 1994, p. 120):

> [F]rom the point of view of mathematics education, what is interesting are [sic] exactly these 'accelerations and regressions' and 'epistemological gaps', as well as 'epistemological obstacles' and difficulties because it is assumed that to learn is to overcome a difficulty. That an equilibrium has to be finally attained — this [sic] is taken as a banality; the problem is that without first destabilizing the student's cognitive structures no process of equilibration will ever occur, i.e., no learning of something radically new will ever occur. (Sierpinska, 1994, p. 121)

Sierpinska argues that in order to improve students' mathematical understanding, teaching should focus on interventions that help students overcome epistemological obstacles.

> So we must introduce the students into new problem situations and expect all kinds of difficulties, misunderstandings and obstacles to emerge and it is our main task as teachers to help the students in overcoming these, in becoming aware of the differences; then the students will perhaps be able to make the necessary reorganizations. (Sierpinska, 1994, p. 122)

Below, are specific types of knowledge that Sierpinska sees as impacting on students' mathematical understanding.

Epistemological Obstacles

Sierpinska asserts that it is through the examination of students' acts of understanding that we can interpret the thinking processes and epistemological obstacles involved in students' construction of meaning regarding mathematical concepts. Sierpinska uses Bachelard's (1975, 1983) notion of epistemological obstacles to describe the learner's unconscious ways of knowing or understanding that constrain his or her ability to think about mathematical concepts in general, elaborated, or more abstract ways. Sierpinska makes the point that although specific methods of measuring students' acts of understanding need to be developed, strategies for teachers to engage students in confronting and overcoming epistemological obstacles also need to be generated:

> [I]nstead of trying to replace the students' 'wrong' knowledge by the 'correct' one, the teacher's effort should be invested into negotiations of meanings with the students, invention of special challenging problems in which a student would experience a mental conflict that would bring to his or her awareness that his or her way of understanding is probably not the only possible one, that it is not universal. (Sierpinska, 1994, xii)

Furthermore, the partial ordering of a learner's acts of understanding would enable the student's depth of mathematical understanding to be compared against criteria and it could be measured in terms of the number and quality of the acts of understanding demonstrated. As well, the number of epistemological obstacles an individual may need to overcome could then be identified.

Sierpinska argues that unlike intuitive knowledge that she describes as irresistible and certain, rational knowledge in mathematics is acquired through the individual's exercise of rigour and attention. Interrogating and synthesizing perspectives of understanding from Locke (1985), Dewey (1988), and Hoyles (1986), Sierpinska generates four categories or acts of conceptual mathematical understanding that she deems necessary for students to experience and use in their studies of mathematics. They are as follows: *identification*, *discrimination*, *generalization*, and *synthesis*. She elaborates on these categories by integrating Vygotsky's (1987) theory of intellectual operations. By doing so, Sierpinska forms a more detailed framework that provides descriptions for the acts and processes involved in students' development of mathematical concepts and the types of epistemological obstacles that may occur.

> The genesis of concepts in a child, according to Vygotsk[y], is the genesis of his or her intellectual operations such as generalization, identification of features of objects, their comparison and differentiation, and synthesis of thoughts in the form of systems. The very same operations lie at the foundations of understanding.... The various genetic forms of these operations, discovered and described by Vygotsk[y], seemed to provide, almost immediately, the possible genetic forms of understanding. Moreover, the theory can be used to explain some of the curious ways in which students understand mathematical notions, and why, at certain stages of their construction of these notions, they simply cannot understand in a different or more elaborate or more abstract way. (Sierpinska, 1994, p. 142-143)

Sierpinska distinguishes two key tenets within this idea of epistemological obstacles and how they affect student understanding of mathematics. First, cognition is not an accumulative process but instead, requires the individual's reflection on past mathematical actions in order for their reconstruction of understanding to occur. It is assumed then, that some form of integration and reorganization is required by the individual in order for his or her knowledge or understanding to move from one level to another. The second assumption is that an individual must rebuild fundamental understandings that give rise to different philosophical considerations in order to overcome an epistemological obstacle. With this process of rebuilding, Sierpinska adds that new knowledge can give rise to future epistemological obstacles through our awareness that an obstacle or obstacles exist in our mathematical understanding or, as a result of the resolution of differences. In this manner, the author describes obstacles as both positive and negative: positive in the sense that we are able to overcome them and negative in the sense that we acquire them.

> [W]e must note that something (a belief, a scheme of thinking) functions as an obstacle often only because either one is unaware of it, or because one does not question it, treating it as dogma. Overcoming an obstacle does not mean switching to another system of beliefs or another persistent and believed universal scheme of thinking but rather in changing the status of these things to 'one possible way of seeing things', 'one possible attitude', or 'a locally valid method of approaching problems' etc. (Sierpinska, 1994, p. 125)

Sierpinska makes it clear to the reader that, unlike Vygotsky's genetic forms of intellectual operations that are chronologically developmental stages, she distinguishes the four categories as co-existing with one another and to be thought of as stages that one progresses through in childhood and adulthood. So even if an adolescent or an adult was confronted with a new mathematical concept, it is likely that the individual would be working with a low level of conceptual understanding of generalization and synthesis or perhaps with a vague discrimination between the relevant and the irrelevant features of that particular concept. Moreover, Sierpinska (1994) makes the argument that:

> It seems that one cannot sensibly speak of epistemological obstacles in children before they reach the age of conceptual thinking. Things went easier with the younger children because they did not have to *overcome* epistemological obstacles. The epistemological obstacles still remained to be *constructed*. (p. 158)

MATHEMATICAL UNDERSTANDING AS A MENTAL PROCESS OF IDENTIFICATION AND DISCRIMINATION

When an individual begins to identify features of objects and distinguishes them as either more or less significant in view of some generalization, this is considered to be a more elaborate form of mental operation. This is illustrated in the following example:

> ... at some point in the process of understanding the topic of equations at the high-school level, the student must identify the simultaneous occurrence of variables and the equal sign as features characteristic of equations before he or she starts to conceptually think of equations as equality conditions on variables. (Sierpinska, 1994, p. 151)

Chain-complexes

Sierpinska uses Vygotsky's term, "chain-complexes" to describe naive generalizations that precede one's development of conceptual mathematical understanding. This type of understanding can be observed in settings that involve actions of sorting or categorizing. Sierpinska (1994) explains that a chain-complex occurs when "a child... adding objects or pictures of objects to a given model, focuses on the last object added and is satisfied with any link between the new object and this last one, disrespectful of any contradiction that may occur with regard to the previously added objects" (p. 147). She uses an example from Vygotsky's research to elucidate this for the reader:

> ... the child may select several objects having corners or angles when a yellow triangle is presented as a model. Then, at some point, a blue object is selected and we find that the child subsequently begins to select other blue objects that may be circles or semicircles. The child then moves on to a new feature and begins to select more circular objects. In the formation of the chained complex, we find these kinds of transitions from one feature to another (Vygotsky, 1987, p. 139)

A second characteristic of chain-complexes is that they usually take place when an individual is developing an understanding for a mathematical concept that involves the notion of equity. In this case, it is not that the individual considers all the attributes of an object as being of equal significance but rather, that the individual is not able to stay focused on one particular feature for any considerable length of time. "At one moment it can be, for example, the colour, at another, the shape" (Sierpinska, 1994, p. 149). Here it is not possible for a student to abstract common features identified from different contexts or to synthesize them into a mathematical concept because the individuals' processes of understanding and the actual object of their understanding is constantly undergoing change.

Pseudo-concept of Generalization

When an individual becomes "aware of the non-essentiality of some assumption, or of the possibility of extending the range of applications" (Sierpinska, 1990, p. 150), this act of mathematical understanding is described by Sierpinska as *generalization*. An obstacle that may occur within this category is a *pseudo-concept of generalization* (Sierpinska, 1993, Sierpinska & Viwegier, 1989), such as when a child identifies geometric shapes according to arbitrary colours so that any object resembling green pattern blocks would be considered squares. This is taken to be a pseudo-concept of generalization because the child makes a generalization but not a mathematical one; for example, that all squares are four-sided polygons.

This epistemological obstacle differs from that of a chain-complex because the learner's way of understanding serves as a more general manner of thinking about mathematical concepts and does not change from situation to situation. Furthermore, due to the pervasive nature of epistemological obstacles, they are not easily abandoned or replaced without considerable reorganization of one's mathematical understanding.

Abstraction

When one is able to maintain one's thinking about the same single feature in order to move beyond complexization (Vygotsky, 1987) and towards the stage of generalization, one also moves closer towards what Sierpinska refers to as true conceptualization. This is preceded by an intermediary phase that she identifies as *potential concepts*. Potential concepts are considered as such because it is possible for the individual to develop an abstract understanding of a mathematical concept once the individual abstracts the underlying idea or ideas at the core of the concrete, factual, or contextual situations.

Conceptual Thinking as Synthesis

The formation of a mathematical concept requires the individual to synthesize features of that concept into a coherent whole. To do so implies that the student constructs mathematical relations between two or more properties, facts, or objects. Sierpinska's research into epistemological obstacles and the role these play in students' struggle to generate and understand mathematical concepts makes it clear to the reader that students cannot achieve or demonstrate conceptual, mathematical thinking through methods that assume learning by telling. Moreover:

> Concepts cannot be given to the child, ready made, in the verbalized form or symbolic representation. The child has to construct them as generalizations of his or her previous generalizations and it is quite natural that the adolescent's first concepts may bear little resemblance to the fully fledge ones developed by generalizations made by mathematicians in their adult, mature, and often genius lives. And thus they become obstacles to understanding the theories. (Sierpinska, 1994, p. 159)

MATHEMATICAL UNDERSTANDING AS ACTIONS, PROCESSES, OBJECTS, AND SCHEMAS VIA REFLECTIVE ABSTRACTION AND GENETIC DECOMPOSITION

Actions

Dubinsky and his collaborators explain mathematical actions to be "any physical or mental transformation of [mathematical] objects to obtain other [mathematical] objects" (Cottrill et al., 1996, p. 171). An example of a physical action could be a student recording or manipulating a mathematical calculation onto paper. A mental action on the other hand, might be a student recalling some mathematical fact such as 6 + 6 = 12, from memory. These actions may consist of one response or a sequence of connected responses that occur when an individual reacts to a perceived external event. Furthermore, it is when the individual reflects on his or her mathematical actions that the actions become a process.

Processes

Cottrill et al. (1996) define a mathematical process to be:

> ... a transformation of an object (or objects) that has the important characteristic that the individual is in control of. . . in the sense that he or she is able to describe, or reflect on, all of the steps in the transformation without necessarily performing them. (p. 171)

Once constructed, a process can be manipulated and combined with other mathematical processes. For example, once an individual understands that "three add four makes seven", this understanding can be reversed and connected to the process of subtraction, "seven take away four makes three". It is these manipulations together with the individual's reflecting on the mathematics at hand that give rise to new processes and can foster the development of relationships between other process constructs to form a schema or, a mathematical object.

Objects

Mathematical objects are described as being "constructed through the encapsulation of a process. This encapsulation is achieved when the individual becomes aware of the totality of the process, realizes that transformations can act on it, and is able to construct such transformation" (Cottrill et al., 1996, p. 171). The student is able to move his or her thinking flexibly back and forth between objects and processes of a mathematical idea. Mathematical objects exist as dense and symbolic schemas with which an individual is able to respond to many different contexts by de-encapsulating a mathematical concept in order to retrieve the appropriate processes or actions.

Schemas

Schemas, as mentioned above, are conceived as coherent mental networks made up of mathematical representations of actions, processes, and objects. Once formed, these networks may also interrelate with other schemas. Moreover, it is through continual, reflective constructing and reconstructing of mathematical schemas that an individual is able to make sense of and deal with problematic situations by modifying or developing new mathematical processes, objects, or schemas.

The Role of Reflective Abstraction in Mathematical Understanding

Dubinsky regards Piaget's notion of reflective abstraction as playing a critical role in the development of students' mathematical thinking. Moreover, Dubinsky supports Piaget's (1985) view that "first... reflective abstraction has no absolute beginning structure and second, that continues up on through higher mathematics" (Beth & Piaget, 1966, p. 203-208). By keeping in one's mind the APOS model (see Figure 6) while reading the following descriptions regarding the different forms of reflective abstraction, one is able to understand how these types of mathematical abstraction play integral parts in an individual's movement from one stage to the next in developing their mathematical thinking from actions to processes, operations, and schemas.

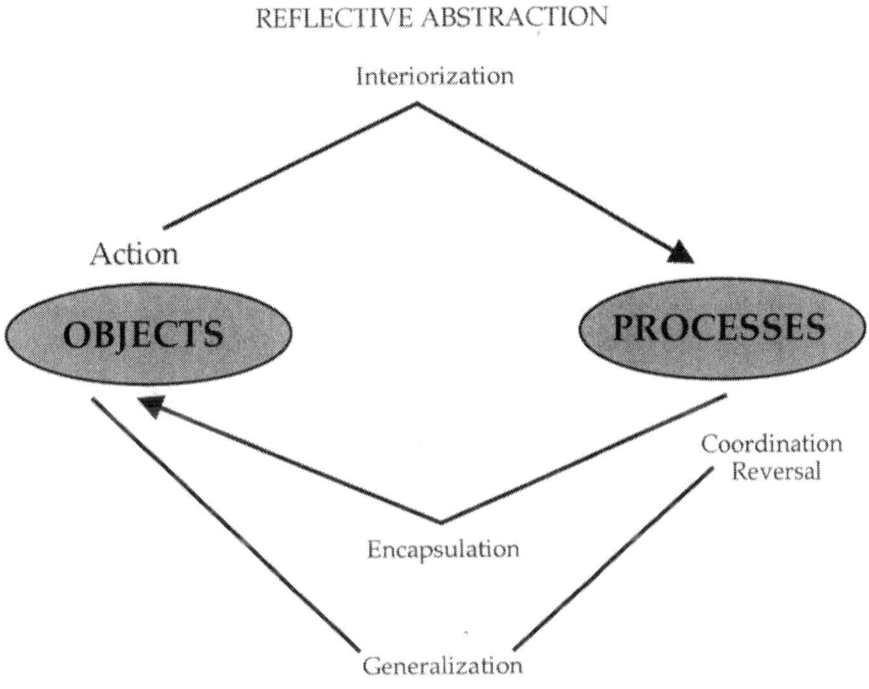

Figure 6. APOS model for the cyclical nature of mathematical actions, processes, objects, and schemas (based on Dubinsky, 1992).

In contrast to empirical and pseudo-empirical forms of abstraction[6], Dubinsky considers reflective abstraction to be the most sophisticated. He

[6] Based on Beth and Piaget (1966), Dubinsky defines empirical abstraction as the least advanced form of abstract thinking; as one's ability to make generalizations about common properties of a collection of objects such as, "All of the blocks are blue." Pseudo-empirical abstraction is considered to be a more advanced form of thinking than empirical abstraction but not as sophisticated as reflective abstraction. This type of internal construction enables an individual to sort out mathematical properties that are being acted out on a set of objects. For example, a person aligning two sets of objects demonstrates that that person has an understanding of one-to-one correspondence among the set of objects.

views reflective abstraction to be necessary for advanced mathematical understanding because it is the process by which students are able to internally coordinate their mathematical actions and form mental mathematical generalizations about external objects or events. Within reflective abstraction, Dubinsky further distinguishes five specific stages of thinking that enable deep mathematical understanding: *interiorization, coordination, encapsulation, generalization,* and *reversal* (see Figure 6).

Interiorization occurs when the learner consciously reflects on mathematical actions and combines them with other actions. Through reflection and *coordinating* two or more mathematical processes together, the individual is able to construct new mathematical objects through *encapsulating* or converting the process(es) into a concept. Once an abstract mathematical object or concept is formed, the learner is able to *generalize* this knowledge and apply it to many different contexts. Further still, this understanding of relationships that exist among mathematical processes, objects, or schemas allows the learner to think flexibly and thus, *reverse* their thinking. For instance, with the understanding of multiplication as the complementary operation of division, it is possible to think about $4 \times 3 = 12$ as the reverse or opposite expression of $12 \div 3 = 4$.

Genetic Decomposition

Dubinsky and his collaborators (Cottrill et al.) propose *genetic decomposition* as a means by which effective teaching methods can foster students' mathematical understanding as framed by the APOS model. Genetic decomposition attempts to identify the elements of thinking that are necessary for students' construction of schemas regarding a particular math concept of say, *limit*. The authors characterize genetic decomposition as a cyclical tool that begins with an analysis of the mathematics for a particular concept and then compares it with students' understanding in order to develop a specific sequence for the learning of the mathematical concept. According to Dubinsky, before any instruction takes place, analytic decomposition is necessary. This approach involves breaking down into smaller chunks and then sequencing mathematical problems into specific steps so that students will be led to construct the particular concept (see Figure 7).

Methods for instruction are then designed and implemented, and observations are collected regarding the mathematical activities of the students. These observations are compared against the first genetic decomposition and necessary revisions regarding the sequencing of the mathematical problems or instructional strategies are made. The entire cycle of decomposition, implementation, and comparison is then repeated until no more revisions are needed (see Figure 8).

Preliminary Genetic Decomposition

Our description of what might occur is organized in six steps that occur only very roughly in the given order and with a great deal of "backing and filling" as the student constructs the concept of limit.

1. The action of evaluating the function f at a few points, each successive point closer to a than was the previous point.
2. Interiorization of the action of Step 1 to a single process in which $f(x)$ approaches L as x approaches a.
3. Encapsulate the process of Step 2 so that, for example, in talking about combination properties of limits, the limit process becomes an object to which actions (e.g., determine if a certain property holds) can be applied.
4. Reconstruct the process of Step 2 in terms of intervals and inequalities. This is done by introducing numerical estimates of the closeness of approach, in symbols, $0 < |x - a| < \delta$ and $|f(x) - L| < \epsilon$.
5. Apply a quantification schema to connect the reconstructed process of the previous step to obtain the formal definition of limit. As we indicated in our comments on the literature, applying this definition is a process in which one imagines iterating through all positive numbers and, for each one called ϵ, visiting every positive number, calling each δ this time, considering each value, called x in the appropriate interval, and checking the inequalities. The implication and the quantification lead to a decision as to whether the definition is satisfied.
6. A completed ϵ-δ conception applied to specific situations.

Figure 7. Example of the genetic decomposition of the concept of limit (as cited in Cottrill et al., 1996, p. 174-17).

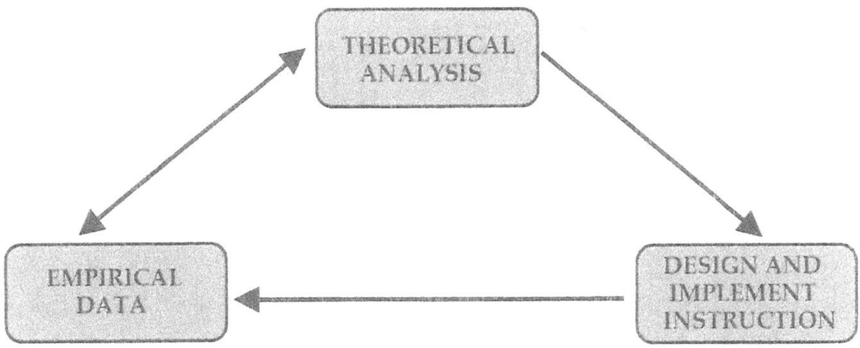

Figure 8. Diagram of the cycle regarding the process of genetic decomposition, its implementation, and its revision (based on Cottrill et al., 1996).

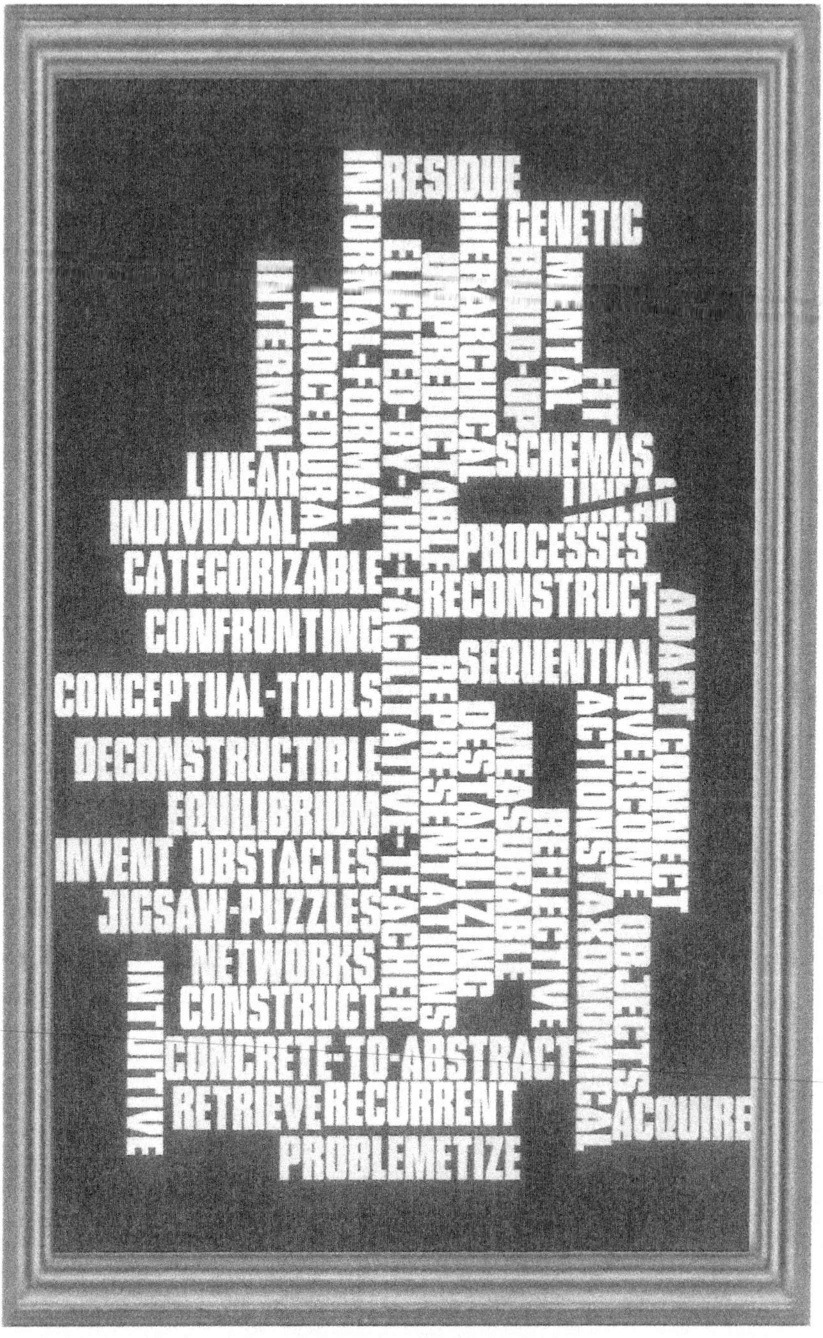
RESIDUE
GENETIC
INFORMAL-FORMAL
ELICITED-BY-THE-FACILITATIVE-TEACHER
UNPREDICTABLE
HIERARCHICAL
BUILD-UP
MENTAL
FIT
PROCEDURAL
INTERNAL
LINEAR
SCHEMAS
INDIVIDUAL
PROCESSES
CATEGORIZABLE
RECONSTRUCT
CONFRONTING
ADAPT
SEQUENTIAL
CONCEPTUAL-TOOLS
REPRESENTATIONS
DESTABILIZING
MEASURABLE
ACTIONS
OVERCOME
CONNECT
DECONSTRUCTIBLE
EQUILIBRIUM
REFLECTIVE
INVENT
OBSTACLES
JIGSAW-PUZZLES
TAXONOMICAL
OBJECTS
NETWORKS
CONSTRUCT
INTUITIVE
CONCRETE-TO-ABSTRACT
RETRIEVE
RECURRENT
ACQUIRE
PROBLEMETIZE

Emphasizing the Dynamic Nature

of Mathematical Understanding

VISUAL AND ANALYTIC STRATEGIES AS INTERRELATED QUALITIES OF MATHEMATICAL UNDERSTANDING

Taking a different approach from that of investigating mathematical understanding through the development of stages and sequences, Zazkis, Dubinsky, and Dautermann's (1996) study examines the possible relationship(s) between the visual and analytic strategies that undergraduate students employed in solving algebraic problems. The students in this study were given mathematical problems that could be solved using either a *visual* (see Figure 9) or *analytic* approach (see Figure 10). In defining these two different strategies, Zazkis et al. describe

> Our specific situation deals with the dihedral group of order four, denoted D_4, and we consider students' thinking about two problems: List the elements of this group, and calculate the products, according to the group operation, of pairs of elements. We chose to observe students working with these D_4 problems because (a) each of the interpretations described below represents roughly the same level of mathematical sophistication, (b) both processes are simple enough to be carried out quickly and are therefore manageable during a clinical interview, and (c) the situation itself is complex enough to bring out distinctions in the students' understanding.
>
> The group D_4 can be modeled in two ways. The approach that we take to be highly related to visual thinking is expressed in terms of the symmetries of a square. In this view, the elements of the group are the four rotations of the square around its center — in 0, 90, 180, and 270 degrees, together with four reflections or "flips" (across lines connecting the midpoints of the opposite sides and the two diagonals). The group operation between two symmetries consists of performing one symmetry on a square and then performing the other on the result. Using this approach, a mathematics student might employ a physical model of the square to achieve an understanding of its various rotations and flips. [Below] illustrates this method of calculating the product of two symmetries. The student performs a vertical flip followed by a 90-degree clockwise rotation to arrive at the reflection or flip across the diagonal with positive slope. (Zazkis et al., 1996, p. 435 - 436).

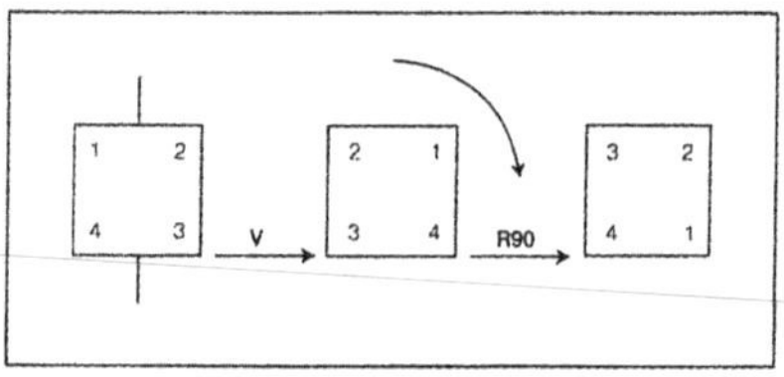

Vertical flip followed by a 90-degree right rotation.

Figure 9. Sample algebraic problem and visual solution (Zazkis et al., 1996).

visualization as occurring when one forms a relationship between what one sees in one's mind as a mental construct and that which is experienced through one's senses in the physical environment. Analytic thinking is characterized as being:

> any mental manipulation of objects or processes with or without the aid of symbols. [For example,] a biologist who analyzes the nature of a plant through decomposing it into its parts, as well as thinking about the relationships among those parts and synthesizing them into various other wholes such as leaves, flowers, and seeds. Thus we include the naming of parts in our view of analysis, but we also include intellectualizing them into various new wholes. (Zazkis et al., 1996, p. 442)

A second approach to D_4, which we take to be more representative of analytic thinking, expresses the group in terms of permutations of four objects. The group operation in this case consists of applying a specific algorithm to multiply these objects and produce a permutation product. Thus a vertical flip of the square might be represented by the permutation

$$\begin{pmatrix}1234\\2143\end{pmatrix}$$

and a 90-degree rotation by the permutation

$$\begin{pmatrix}1234\\2341\end{pmatrix}.$$

After multiplying these permutation we get

$$\begin{pmatrix}1234\\3214\end{pmatrix},$$

Figure 10. Zazkis et al.'s example of analytic solution for the same algebraic problem. (as cited in Zazkis et al., 1996, p. 436.)

The authors' definition of mathematical analysis also includes the five previously mentioned forms of reflective abstractions[1] that enable the individual to construct mental representations.

Zazkis et al.'s research explores previous claims that attribute visual strategies as being less sophisticated (Eisenberg & Dreyfus, 1991; Gollwitzer, 1991; Presmeg, 1986a; Vinner, 1989) and even restrictive to students' mathematical abilities in connecting visual representations to symbolic forms of mathematics (Kruteskii, 1976; Presmeg, 1986b). Interestingly, the authors found that the students did not use one or the other of visual or analytic strategies but employed a combination of the two approaches. Consequently, they developed an alternative model that does not dichotomize visualization and analysis but puts forth the idea of the

[1] Interiorization, coordination, encapsulation, generalization, and reversal.

two as interrelated. Thus, visual and analytic thinking are "interacting and mutually supporting modes of thinking, rather than as two sides of a coin or as a dichotomy or continuum" (Zazkis et al., 1996, p. 454). The authors' visualizer-analyzer (VA) model conceptualizes the two problem solving approaches to be interdependent and challenges Piaget's (1977) claim that "some people are particularly visual, others mainly motor, auditory, etc." (p. 684). Moreover, Zazkis et al. align their view to be more in keeping with Clements' (1982a) work that explores learners as not only being visualizers and verbalizers but also, *mixers* — those individuals who "do not have a tendency one way or the other" (Clements, 1982b, p. 34).

Instead of describing student visualization and analysis as being located on opposite ends of a continuum, the VA model (see Figure 11) makes the assumption that:

> [i]t could be that a preference for, and difficulties with, visualization (Bishop, 1986; Goldenberg, 1991; Tall, 1991; Vinner, 1989) is no more than an individual's tendency to dwell on one side or another of the triangle, for example, when communicating her or his thinking an individual might be more comfortable drawing pictures or writing formulas, but that does not change the fact that he or she needs analytic thinking in determining what to draw, and he or she eventually constructs a rich mental picture that determines what symbols to write. (Zazkis et al., 1996, p. 453)

As Zazkis et al. assume visualization and analysis to be intertwined and therefore inseparable, they state that it is impossible to make claims that categorize students' mathematical thinking or prioritize one method over the other. This model serves to conceptualize visual and analytic thinking as being equally important in students' development of fluid and rich mathematical understanding.

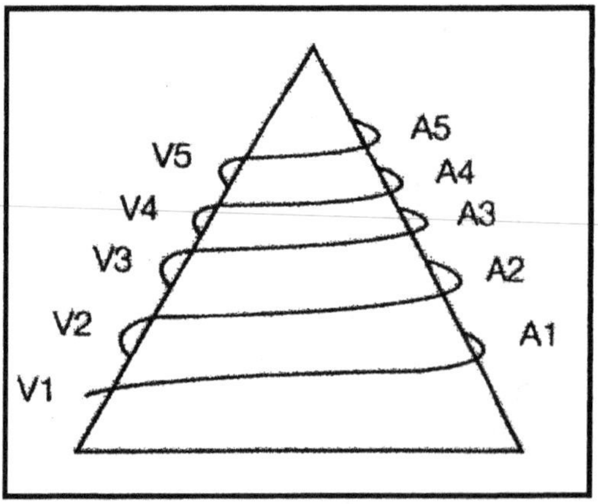

Figure 11. Visualization/Analysis Model (Zazkis et al., 1996).

~~LINEAR~~ AND ~~HIERARCHICAL~~ VIEWS OF MATHEMATICAL UNDERSTANDING

Schoenfeld argues against linear, hierarchical frameworks that characterize learners' mathematical knowledge as existing as a series of stages that begin with naive understandings and move progressively towards formal mathematical knowledge. This is because the structure of such frameworks does not highlight the fragmented and unstable nature of learners' knowledge structures. Second, Schoenfeld does not conceptualize mathematical understanding as being unidirectional but views it as occurring through a back and forth or bidirectional manner. Thirdly and in similar vein to Hiebert (1989) and Pimm (1987), Schoenfeld (1991) advocates for models that focus on learners' mathematical understanding as continuous and not those that "trap" it in structures of linearity. Thus, by describing mathematical understanding as being circular in nature emphasizes mathematical ways of knowing as being constantly reintegrated and giving rise to more complex forms of knowledge.

Moreover, in order for educators to gain an understanding regarding learners' development of mathematical concepts, teachers must focus on examining the dynamics of learners' mathematical actions as well as how they evolve through language and social interactions (Schoenfeld, 1996). Mathematics as Schoenfeld describes it, is all about deep, connected understandings; that is, understandings that occur not only within an individual's mind but also collaboratively and socially through the interacting with others:

> One often thinks of the stereotype, the isolate mathematician alone in his office, struggling to prove a new theorem. This is certainly a part of mathematics, but there is a social aspect of it as well, an aspect that Diaconis captured perfectly. Coming to grips with mathematics involves "talking and explaining, false starts, and the interaction of personalities." All of it, not the least of which is the challenge of the false starts, is indeed a great joy. (Schoenfeld, 1991, p. 328)

Schoenfeld (1992) applies Ryle's (1949) description to distinguish instrumental knowledge as being "knowledge that" and relational knowledge as "knowing how". He also supports Hiebert's (1985) view that there is not a distinct line or boundary that separates these two forms of mathematical knowledge but rather, each informs and gives rise to the other.

According to Schoenfeld, mathematical understanding that is to be conceptualized as fluid and dynamic cannot be explored through models that assume mathematical development to be a monitonic process that entails the adding on of more knowledge to one's knowledge base. Instead, research that aims to describe mathematical understanding as fluid and dynamic demands multiple perspectives that not only address specific issues of mathematical understanding but also connect individual theories to larger realms within mathematics education:

> [W] e need focus and pluralism, and an occasional step back to look at the big picture.... there should be broad diversity in what we look at, and the methods we use to do the looking — I don't believe unified theories or methodologies are around the corner — ....We need to work on our descriptions both of the forest and of the trees within. (Schoenfeld, 1989a, p. 116)

A Model for Analyzing Students' Mathematical Understanding

Schoenfeld's (1989a) video analysis examines students' conceptual understanding of mathematics through different levels of detail as it occurs during problem solving situations. His research explores the possibilities of computer-based learning for enhancing mathematical understanding as it pertains to students' graphing of straight lines (see Figure 12).

Schoenfeld's analytic model enables comparisons to be made between students' mathematical structures (the right-hand column) and preestablished mathematical forms (the middle column) as well as defining a student's level of complexity with respect to their conceptual schema(s). As seen in Figure 12, the left-hand column represents the particular lens through which the researcher is examining student interactions; that is, the macro level focuses on the student's mathematical schemas, the middle level examines the entailments of larger schemas, the micro level moves closer in to the connections associated with the entailments of schemas, and the fourth level zooms in to the contexts that give rise to the student's micro level of understanding. Schoenfeld's leveled structure of analysis enables the researcher not only to deconstruct mathematical concepts and students' conceptual understanding but also to tease out the relationships that exist amongst them.

Consequently, Schoenfeld conceives deep mathematical understanding as being that which allows for flexibility and proficiency because of its rich, well developed microlevel of a knowledge base. Unstable mathematical understandings and misconceptions on the other hand are taken to be a result of an individual's ill-grounded connections at the micro level. Furthermore, Schoenfeld, like R. Davis (1992) views formal mathematics as a product or residue of well-connected mathematical understanding. He emphasizes these points below:

> If you understand how things fit together in mathematics, there is very little to memorize. That is, the important thing in mathematics is seeing the connections, seeing what makes things tick and how they fit together. Doing the mathematics is putting together the connections and making sense of the structure. Writing down the results — the formal statements that codify your understanding — is the end product, rather than the starting place. (Schoenfeld, 1991, p. 328)

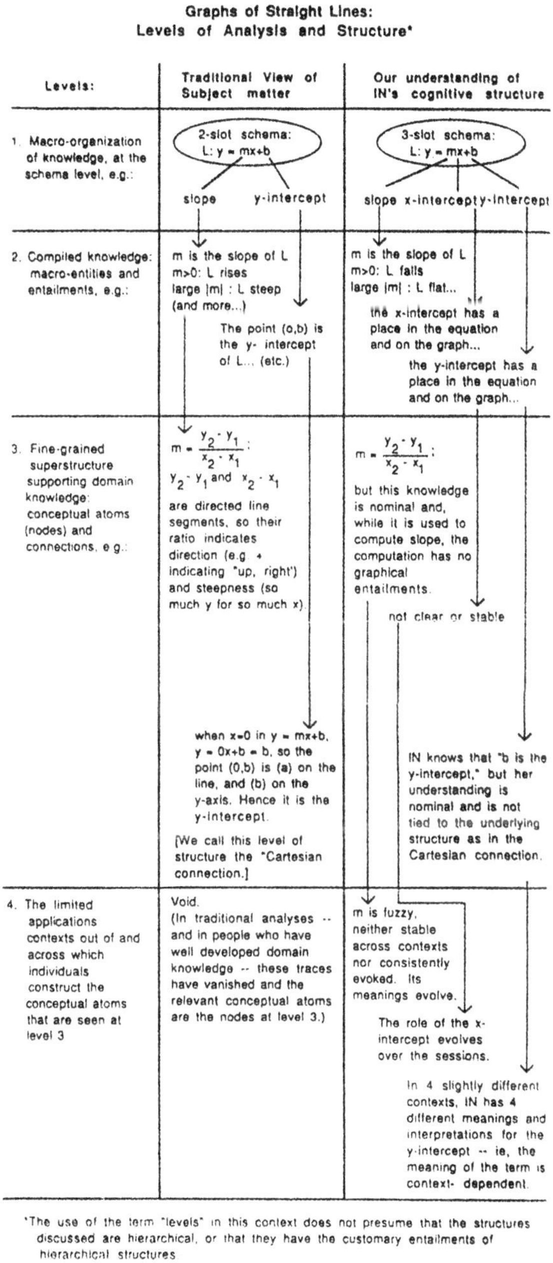

Figure 12. An example demonstrating Schoenfeld's (1989a) leveled analysis of the graphing of straight lines and the comparison between that of established mathematics and that of student mathematical understanding.

Problem Solving for the Development of Mathematical Understanding

Schoenfeld (1992) further elaborates on his analytic model by taking a step closer and looking at individuals' mathematical knowledge bases. Here he examines what relevant information that students draw on during mathematical problem solving and the ways in which they retrieve and employ this information. For example, given the problem (see Figure 13), a student's mathematical understanding can be assessed against the table below (see Figure 14):

> You are given two intersecting straight lines and a point P marked on one of them, as in the figure below. Show how to construct, using a straightedge and compass, a circle that is tangent to both lines and that as the point P as its point of tangency to one of the lines.

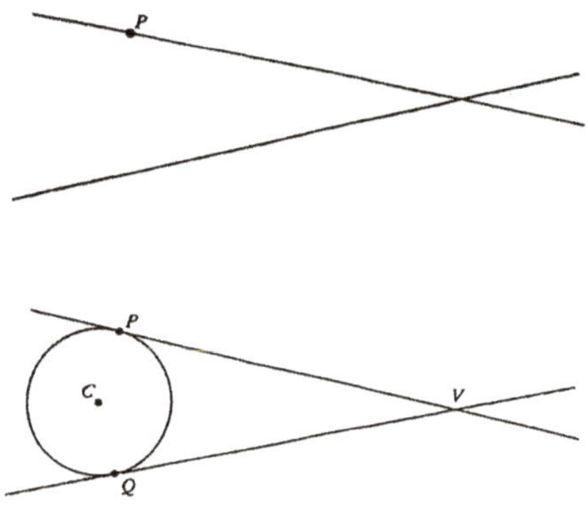

Figure 13. Example of a mathematical problem and its solution used to analyze student problem solving action. (as cited in Schoenfeld, 1992, p. 349-350).

Degree of Knowledge of facts and procedures

Does the student:

Degree of Knowledge	facts	procedures
a. know nothing about b. know about the existence of, but nothing about the details of c. partially recall or suspect the details, but with little certainty d. confidently believe	the tangent to a circle is perpendicular to the radius drawn to the point of tangency (true) any two constructible loci suffice to determine the location of a point (true with qualifications) the center of an inscribed circle in a triangle lies at the intersection of the medians (false)	a (correct) procedure for bisecting an angle a (correct) procedure for dropping a perpendicular to a line from a point an (incorrect) procedure for erecting a perpendicular to a line through a given point on that line

Figure 14. Partial inventory of an individual's resources for working out the construction problem as described in Figure 13 (Schoenfeld, 1992).

Here, informal knowledge is defined as the knowledge that a student brings to bear on a particular problem such as the student's mathematical intuitions whereas more formal knowledge consists of mathematical facts, definitions, or algorithmic procedures. The ways in which these forms of knowledge are expressed may vary depending on the individual's confidence or certainty of them.

The Structure of Memory: Access to Resources

Schoenfeld proposes another model that outlines his conceptualization of our memory system and how the contents of memory are organized, accessed, and processed in a sequential yet somewhat circular manner (see Figure 15). Schoenfeld explains that visual, auditory, and tactile information is received through what he calls as sensory buffers or, short-term memory. Short-term memory is described as the location where the thinking gets done. If sensory information is attended to within one's short-term memory, then it is converted into forms that are further developed through the working and long-term memory systems. Working memory is different from that of short and long term memory because not only does one's working memory take in information from these other two sites but it is also where metalevel processes occur and enable one to construct mental representations. In addition to this, within one's working memory, one is able to structure planning, monitor, and evaluate one's mathematical actions. It is this activity that takes place in our working memory that enables us to externalize our mathematical thinking through various

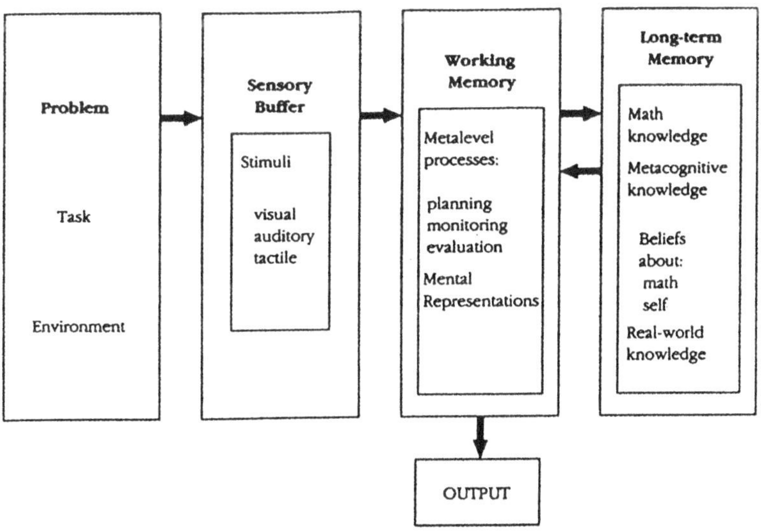

Figure 15. Schoenfeld's (1992) conceptualization of the structure of memory.

physical, written, or verbal forms of expression. Long-term memory system is considered as a "permanent knowledge repository" (Schoenfeld, 1992, p. 351). It is a neural network in which mathematical knowledge functions as nodes and relational knowledge as the strands that connect these nodes together to form the network.

Schoenfeld's (1985, 1987, 1989b) other studies that examined and compared students' executive or control skills (see Figure 16) to those of expert mathematicians (see Figure 17) revealed that unlike the students who spent much of their time exploring the mathematical problems, mathematicians spent the majority of their problem solving time making sense of the situation, analyzing, and structuring their exploration; that is, in thinking through the situation at hand, mathematicians produced solutions through generating and implementing devised methods that demonstrated a high level of control and perseverance. In the context of mathematical problem solving, *control* refers to the way in which an individual selects goals and subgoals, monitors, revises, and assesses their progress of a problem solving activity. Control also includes how one makes use of and sense of given or found information in attempts to solve a problem. The second managerial strategy, *perseverance* refers to an individual's intuitive, experiential sense in knowing when to continue with and not give up too soon on a chosen strategy or action but also, knowing when to abandon a particular strategy or action and search for a more effective or useful one.

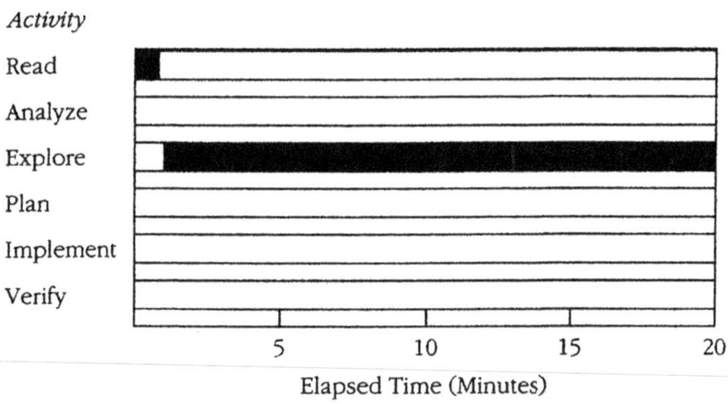

Figure 16. Time-line graph of a typical student attempting to solve a non-standard problem (Schoenfeld, 1992).

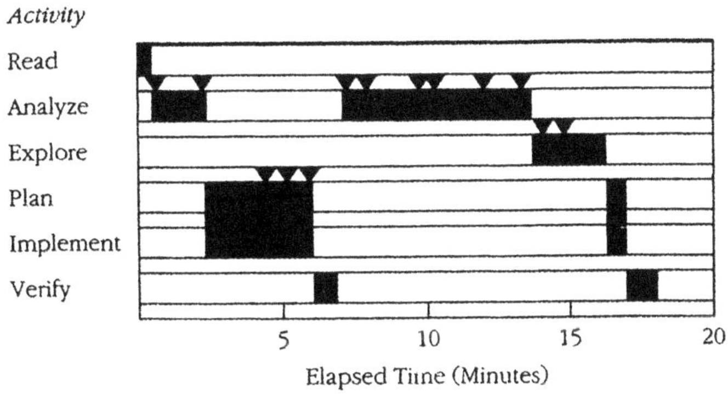

Figure 17. Time-line graph of a mathematician working on a difficult problem (Schoenfeld, 1992).

Based on these observations and implementing instruction that focuses on students' development of control and perseverance in their mathematical thinking (see Figure 18), Schoenfeld's (1989b; 1992) work supports other researchers' (Carraher et. al, 1987; Hart, 1989; Hiebert, 1989; Taplin, 1995) contentions for perseverance and control as two critical qualities necessary for well developed mathematical understanding. Schoenfeld advocates teaching methods that engage students in reflecting and routinely explaining their mathematical actions as well as in providing reasons for why their actions make sense within the given context. In doing so, he states that these managerial skills will then become a natural way of thinking about mathematics and enable more complex mathematical understandings to occur.

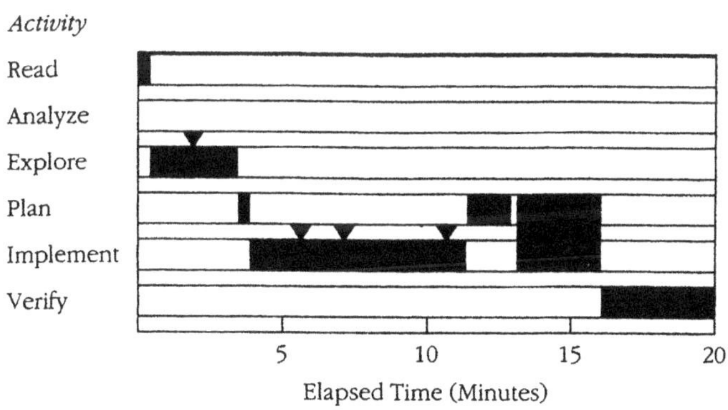

Figure 18. Time-line graph of student's problem solving actions after implementation of instruction that focused on development of metacognitive problem solving skills (Schoenfeld, 1992).

THE METAPHORICAL NATURE OF MATHEMATICAL UNDERSTANDING AND THE PROCESSES BY WHICH UNDERSTANDING OCCURS

Sfard examines mathematical understanding as being rooted in and growing from one's use of conceptual metaphors. Metaphors, she asserts, not only provide us with a means by which to explain our thinking, but they also shape our ways of understanding and knowing mathematics. Sfard expresses this when she speaks of Reddy's (1978) notion of the conduit metaphor and connects it to that of mathematician's conceptions of what it means to understand mathematics:

> Rather than being just tools for a better understanding and memorizing, conceptual metaphors are often the primary source of mathematical concepts. The constitutive role of metaphor has been mentioned explicitly by the mathematicians whom I have interviewed in one of my studies. (Sfard, 1994)
>
> In this way or another, all of them made it clear that without a metaphor, a new concept is not a concept at all. They also repeatedly emphasized the indispensability of the metaphor in the subsequent problem-solving process. (Sfard, 1994, as cited in Sfard, 1997, p. 340)

Sfard (2000) also makes the observation that there are two main types of educational metaphors used in mathematics education that characterize learners' mathematical understanding in two different manners; that is, metaphors that describe mathematical understanding as a process of *acquisition* and metaphors that describe mathematical understanding as developed through *participation*. Acquistionist metaphors are defined by Sfard as views that describe learners' conceptual understanding of mathematics to be a process by which "basic units of knowledge. . . can be accumulated, gradually refined, and combined to form ever richer cognitive structures." (1998, p. 5). Furthermore, Sfard distinguishes that:

> The picture is not much different when we talk about the learner as a person who constructs meaning. This approach, which today seems natural and self-evident, brings to mind the activity of accumulating material goods. The language of "knowledge acquisition" and "concept development" makes us think about the human mind as a container to be filled with certain materials and abut the learner as becoming an owner of these materials. (1998, p. 5)

She contrasts the acquisition metaphor with a participation metaphor by stating that, "the PM [participation metaphor] shifts the focus to the evolving bonds between the individual and others.... Indeed, PM makes salient the dialectic nature of the learning interaction: The whole and the parts affect and inform each other" (1998, p. 5). As well, she notes that unlike the acquisition metaphor that emphasizes mathematical knowledge as a product of learning and teaching, the participation metaphor amplifies mathematical knowing as occurring through ongoing interaction within a mathematical community.

Sfard (1998) recognizes the need for metaphors to be flexible and diverse because "...too great a devotion to one particular metaphor and rejection of all others can lead to theoretical distortions and to undesirable practical consequences" (p. 5):

> We have to accept the fact that the metaphors we use while theorizing may be good enough to fit small areas, but none of them suffice to cover the entire field. In other words, we must learn to satisfy ourselves with only local sense making. A realistic thinker knows he or she has to give up the hope that the little patches of coherence will eventually combine into a consistent global theory. It seems that the sooner we accept the thought that our work is bound to produce patchwork of metaphors rather than a unified, homogeneous theory of learning, the better for us and for those whose lives are likely to be affected by our work. (Sfard, 1998, p. 12)

Sfard does not enter into debates that aim to delineate learning as being conceptualized through either acquistionist or participative metaphors but rather, thinks that we should take the best qualities of both metaphorical ways of thinking and use them not in a divisive manner but in an integrated manner; that is, that discourse should focus on distinguishing contexts in which applications of each approach proves effective. In addition to this, Sfard stresses that even if we wanted to subscribe to framing mathematical understanding as say, solely participatory in nature, due to our cultural embeddedness in acquisitionist language we cannot help but to think acquisitionally, with objects and abstract properties — it is a part of our taken for granted ways of being. Sfard (1997) states that both metaphorical ways of thinking offer qualities that the other cannot and in doing so, argues that "the most powerful research is the one that stands on more than one metaphorical leg" (Sfard, 1998, p. 11) because these metaphors provide tension from which theories can be interrogated.

Sfard identifies one limitation that can occur when only using a participative approach to teaching mathematics; that is, that this way of thinking about mathematical learning can lead to the "complete delegitimatization of instruction that is not problem based or not situated in a real-life context" (1998, p. 11) and "[t]his is difficult, when mathematics at some point exists within the symbolic, abstract realm" (Sfard, 1998, p. 36). Conversely, Sfard explains that when applying a solely acquisitional approach to teaching mathematics together with the assumption that learners build their own conceptual understanding of mathematics, the problem of bridging individual and collective knowledge becomes difficult.

Connecting Mathematical Processes of Knowing with Objects of Knowledge: Operational and Structural Conceptions of Mathematics

Instead of describing mathematical understanding as that which exists as either object or action, Sfard's research (1992) attempts to bridge this dichotomous gap and establishes the need for the co-existence of both mathematical knowing and knowledge; that is, that "an adequate combination of the AM and the PM would bring to the fore the advantages of them" (1998, p. 11). In describing the conceptual development of mathematics, Sfard (1997) characterizes it as being "a zig-zag movement with our conceptual schemes as constituting an "autopoietic system"; that is, a system which is "continually self-producing" (Maturana & Varela, 1987, p. 43). These qualities regarding mathematical understanding are pervasive elements throughout Sfard's research and reflect the value she holds for structural or abstract knowledge and operational or context-bound knowings. Sfard seeks to describe the interrelationships that connect mathematical knowledge and knowing by examining the processes that facilitate the formation of abstract, symbolic, concepts in mathematics.

Sfard's (1991) identifies three hierarchical stages of mathematical conceptions. The stages are referred to as: *interiorization, condensation,* and *reification* (see Figure 19). She defines the interiorization as the stage in which a learner develops an operational concept of the mathematics in order to perform an action on a given problem. For example, "When I fill each of these three boxes up to the top and count the total number of cubes, I can find out how much each container holds". All operational concepts that are formed within the interiorization stage are considered to be context-specific knowings.

Sfard describes the second stage — the condensation stage — as being when learners are able to metaphorically, stand back and begin to reintegrate or make generalizations about their mathematical understandings. It is this middle stage where the interplay between the synthesis of the learner's previous operational mathematical conceptions and a move towards the formation of an abstract, structural concept occurs. Using again, the example of the container, a student might now think, "I filled and counted the number of cubes that each of the three containers held. I wonder if there's a way that I can determine how many cubes the containers hold without having to fill each box and count the cubes one by one?"

The final stage of reification is explained by Sfard as the point at which the learner comprehends the mathematical concept — in this case, the volume of rectangular prisms as an *object* or a thing that is symbolic, dense, and versatile. So, understanding that, "If I want to know the total number of cubes that *any* box can hold, all I need to do is to measure (with cubes) and multiply the length of the box by the width of the box by the height of the box." Hence, by condensing one's knowing of operational, context-specific actions through ongoing analysis, one's mathematical knowledge can become reified into a flexible, structural form.

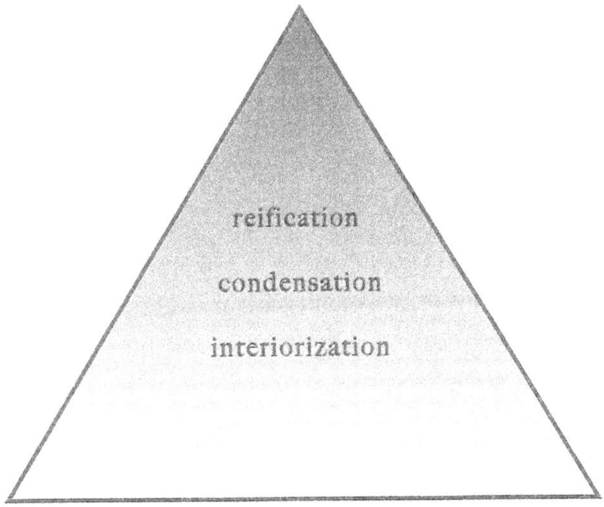

Figure 19. Diagram that characterizes Sfard's (1991) configuration of the three stages of mathematical conception.

The Limitations of Metaphors: Issues of Integration, Overprojection, and Confinement

Sfard points out that although metaphors enable us to think about mathematics in powerfully abstract and symbolic ways, at the same time, they are shaped and limited within the confines of our experiential knowledge. Below are ontological obstacles the author identifies as they pertain to the integration of metaphors, metaphorical overprojection, and metaphorical confinement. In keeping with Sierpinska[2] and R. Davis & Vinner's[3] views, Sfard too considers it necessary for students to overcome these obstacles in order to integrate many different metaphors and develop a sound, stable conceptual knowledge of mathematics.

Sfard explains that one possible reason for students' difficulties with integrating metaphors is due to their inability to allow certain characteristics of the metaphor to fade into the background in order to integrate new qualities that extend their knowledge to new or different mathematical situations. For example, when learning about the concept of division, one must in a sense, forget one's previous understanding that the operation of division when applied to whole numbers, *makes the quotient smaller,* in order to develop an understanding for why division *makes the quotient bigger* when working with fractions and decimals.

Instances when metaphorical overprojection takes place involve situations when an inconsistency lies in the student's mathematical actions:

---

[2] As discussed on p. 95-99.
[3] As discussed on p. 87-88 and p. 89-90.

> Without abandonment of certain characteristics there may be a danger of a logical incompatibility with the new context or with metaphors contributing to the construction of the new concept. Appropriate modifications, however, are sometimes difficult to perform. Certain characteristics, being a vital component of the source notion, would refuse to go. (Sfard, 1997, p. 368)

She provides the following example of a student who divides the factor 'x - 2' into both sides of the equation, $(x-2)^2(x+3)=2(x-2)$. This is considered to be an overprojection of the metaphor, *an equation is a balance*, because when the student performs this operation on both sides of the equation, the student is thinking that equality has been maintained when in actual fact, the root x=2 has been lost. Metaphorical overprojection can also occur when a student tries to integrate two incompatible metaphors such as, *number as quantity* with the concept of complex numbers. If students cannot exclude or in some way forget the quantitative quality of numbers, there results an incompatibility that limits and proves problematic to their conceptual understanding of complex numbers.

Metaphorical confinement as a third ontological obstacle occurs when a student's metaphor is not broad enough to allow for the development of different, related metaphors or mathematical concepts. This form of confinement in mathematical understanding is present when students can only visualize fractions as being *part(s) of a whole;* with this image, their conceptual understanding is confined and because it cannot be opened to fractions existing as *object(s) within a larger group of objects,* or as another way of expressing the divisive action.

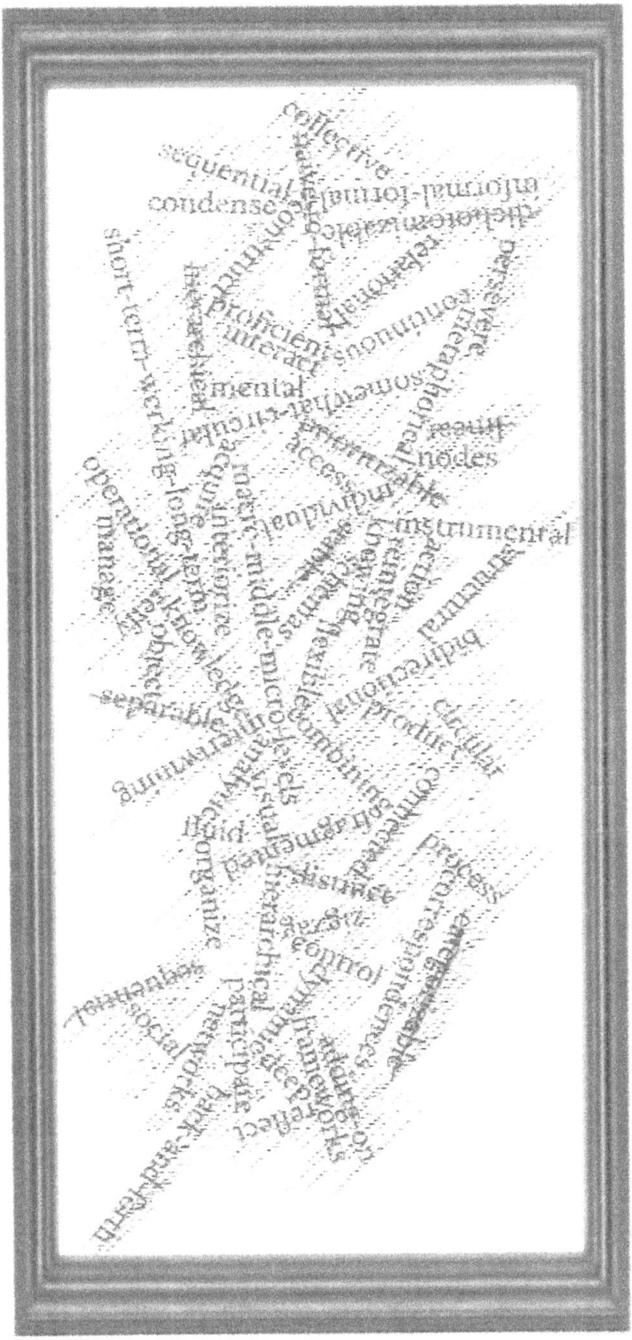
collective
sequential
informal-formal
condense
relational
continuous
proficient
interact
mental
metaphorical
linear
nodes
accessible
individual
instrumental
action
structural
bidirectional
reintegrate
flexible
schemas
product
circular
connected
process
distinct
control
hierarchical
participate
networks
social
back-and-forth
operational
manage
object
separable
knowledge
interiorize
acquire
organize
fluid

Rethinking Mathematical Understanding

from an Enactive Perspective

MATHEMATICAL UNDERSTANDING AS PERSONAL AND PUBLIC FORMS OF COMMUNICATION

In still other works, Sfard shifts her perspective towards an enactive one. Mathematical thinking is regarded by Sfard (2000) as a form of communication; an integral cognitive process which allows individuals not only to express with others how and what they understand about the mathematics at hand but also, mathematical thinking as communication shapes how we individually and collectively make sense of mathematics. She contends that our reasons for communicating are not to establish mathematical objects but rather, mathematical objects are necessary for the conceptual development in our own internal thinking and when interacting with others. Sfard explains that mathematical objects whether physical, verbal, or mental arise as a product of our need to communicate; not the other way around: We do not start with mathematical objects and then communicate, we communicate and through this dialogic process, mathematical objects come into being. In keeping with this, Sfard (2000) makes the assertion:

> I will argue that the claim of the primacy of communication imposes a literal reversal of this relationship: Instead of being merely helpful in constructing and sharing the knowledge of preexisting mathematical objects, communication and its demands must now be regarded as the primary cause for their existence. (p. 4)

For these reasons Sfard's research and the work that she has done with Kieran (Sfard & Kieran, 2001) reveal mathematical communication as having a positive, neutral, or even detrimental impact on students' conceptual understanding. Sfard and, Sfard and Kieran's research closes the gap between our conceptions of students' internal, cognitive thinking as being separate from their interactive and communicative ways of acting. As well, these works bring forth the notion of internal and collective mathematical understanding as being conceived together.

Sfard (1997) supports Maturana and Varela's (1987) view that cognition in its most encompassing sense, co-evolves from our ways of knowing, our actions, and in our individual and collective identities. This is evident when she speaks of mathematical understanding as being shaped and evolving within ourselves and with others. Similarly to that of Gadamer (1989) who described understanding as a conversation — dynamic, unpredictable, and dependent on those participating in the conversation, Sfard (2000) too uses this metaphor when she characterizes mathematical thinking as it occurs when individual students work collectively in a larger group:

> Thinking, like conversation between two people, involves turn taking, asking questions and giving answers, and building each new utterance — whether audible or silent, whether in words or in other symbols — on previous ones in such a manner that all are interconnected in an essential way. (p. 5)

She argues that rather than simply viewing mathematical understanding as being that which exists either in the objective, public realm or in the individual, private realm, we need to also focus on the relationships that emerge between formal mathematics and that which is considered to be informal and embodied, the connections which take place within individual and collective conversations and the ways in which these interactions affect the kinds of mathematical understanding that result.

Mathematical Conceptualization as Complex Circularity

By integrating Sfard's work which illuminates students' mathematical thinking as being circular and complex with Sfard's model of mathematical conception (1991), the latter shifts from that which was linear in structure to one that portrays mathematical understanding to be co-emergent and cyclical (see Figure 20). This integration of theory is possible when we examine Sfard's (2000) definition for *attended* focus. She describes this as being the mathematics or mathematical object that arises as the individual or group's subject of conversation. Here as in Sfard's previous model of concept formation, this can be seen as corresponding to the condensation stage. In both cases, attended focus and condensation stage, there is an

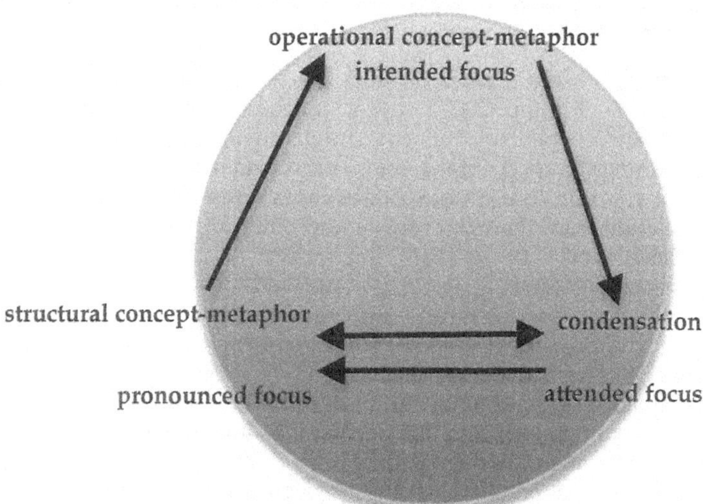

Figure 20. Diagram that integrates Sfard's (1991, 2000) and Sfard and Kieran's (2001) theories regarding mathematical conception and mathematical communication.

interactivity that involves the weaving together of several foci. Secondly the form of mathematics or mathematical thinking that resembles the reification stage can be seen in what Sfard (2000) describes as a *pronounced* focus of conversation; when "the learner[s] can flexibly move back and forth if needed to other realms whereby effective communication mediates these transitions" (p. 33). Thirdly, that which is considered to be the *intended* focus of mathematical communication, specified on an individual level as being each person's interpretation of the pronounced and attended foci, fits with Sfard's operational metaphors located in the interiorization stage.

MATHEMATICAL UNDERSTANDING FROM WITHIN A COMMOGNITIVE FRAMEWORK

As the work of Sfard and, Sfard and Kieran focuses on the participatory and discursive qualities of mathematical learning, it can in a general sense, be viewed as consistent with the theorizing of Cobb and his collaborators (e.g., Gravemeijer, Cobb, Bowers & Whitenack, 2000; Yackel & Cobb, 1996). However, Sfard (2007) argues that because her research, at its core, assumes thinking to be communication, this assumption enables her work to focus on revealing distinct yet complementary aspects of mathematical understanding. She uses the term, *commognitive,* to emphasize the inseparability she conceives between communication and cognition.

Within the commognitive framework, Sfard considers personal understanding as the further sense-making that arises from mathematical engagement with others and allows one to work independently. That said, it is important to keep in mind that Sfard does not view mathematical understanding as a one-way process. She explains that when personal elaborations are taken up by other group members and become part of the group's mathematical activity, individual understanding in this manner, affects what occurs at the collective level through the process of communalization.

Moving into the particulars of this theory while still working from the premise that thinking is communication, Sfard locates *mathematical words, mediators, endorsed narratives,* and *routines* as four discursive sites where individual and collective understanding may be observed. Mathematical words (and terms) possess meanings and applications in the discourse of mathematics. Mediators are artifacts such as diagrams, graphs, and formulae that enable members to distinguish the object(s) of their mathematical talk as well as communicate with one another. Narratives include not only accounts about mathematical objects such as 9 - 5 = 4 but on a meta-level, narratives are concerned with the mathematics itself; for instance, knowing what to do to determine the sum of two values. And, Sfard defines routines as repetitive, established practices that are observable in one's mathematical actions. By definition, commognitive routines map onto the other three sites but in a more general manner.

Thus, transformation at any of these sites signals a change in mathematical understanding whether it be at an individual or collective level. Sfard describes these changes in understanding as being occasioned by commognitive conflict. Here, commognitive conflicts are not taken to be mismatches between the learner's understanding and the external environment but rather, exist as perceived differences in others' manners of communication with whom the learner engages. Sfard asserts that commognitive conflicts are not only critical in furthering learners' mathematical understanding at the object-level but especially important at the meta-level; that is, without commognitive conflicts, there exists no provocation for learners' to change their discursive manners of making sense of the mathematics. Effectively, commognitive conflicts are not obstructions that teachers should prevent learners from encountering or those that need to be overcome by matter of acceptance or refusal; instead, they serve as occasions to make new sense from the mathematical discourses of others.

MATHEMATICAL UNDERSTANDING AS GROWTH: THE PIRIE-KIEREN MODEL

Pirie and Kieren's cognitive mappings of individuals and groups of students identify mathematical understanding as being simultaneously individual and collective, dynamic, occurring on many levels at once, and revealing qualities of transcendence and recursion (see Figure 21 and 22). The authors consider learners to be autopoietic beings who determine

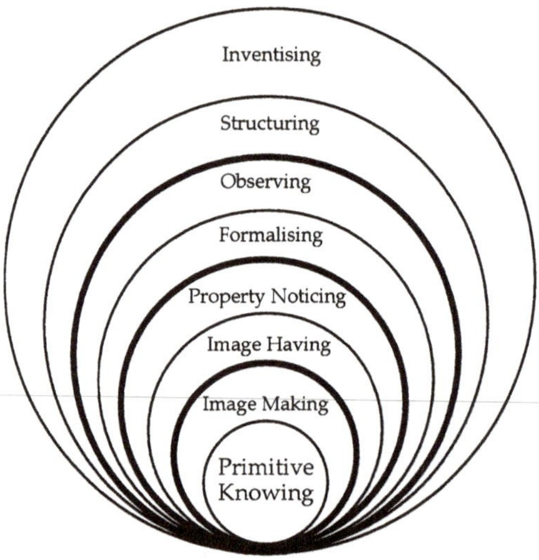

Figure 21. Model of a dynamical theory of the growth of mathematical understanding (adapted from Pirie & Kieren, 1994b).

Definitions of Levels of Mathematical Growth

Primitive Doing or Knowing: all of the knowledge that a learner or group of learners brings to the particular mathematics and from which all new understandings develop.

Image Making: making distinctions in previous knowing and using it in new ways. "Image" not only includes physical and verbal forms, but mental representations as well.

Image Having: using a mental construct about a topic without performing the particular activities that brought it about.

Property Noticing: making note of distinctions, combinations or connections between images, predicting how they might be achieved and recording such relationships.

Formalising: abstracting a method or common quality from the noted properties that are not dependent on meaningful images.

Observing: reflecting on and coordinating formal activity, expressing coordinations such as theorems.

Structuring: explaining or theorizing one's formal observations in terms of a logical structure.

Inventising: breaking away from preconceptions that brought about previous understanding and creating new questions that might grow into a completely different concept.

Other Features of the Model

- **Folding back:** moving to an inner level in order to extend one's current, inadequate understanding when faced with problems at any level.
- **"Don't need" boundaries:** indicated by the model's bold rings; conveys the idea that beyond the boundary one does not need the specific inner understanding that gave rise to the outer knowing.
- Each level beyond primitive knowing is composed of a complementarity of acting and expressing necessary before one is able move to the next level; **acting** encompasses all previous understanding, and **expressing** gives distinct substance to that particular level.

Figure 22. Definitions of terms and features of the Pirie-Kieren model for the growth of mathematical understanding.

what phenomena will be experienced as perturbatory and who specify the ways in which they structure their mathematical thinking (Kieren & Pirie, 1992). As well, Pirie and Kieren argue that mathematical understanding does not occur as a result of student interactions with others or the environment but rather, comes to be through the structural changes within, between, and among learners and the environment (Kieren, Gordon

Calvert, Reid, & Simmt, 1995; Gordon Calvert, 1999; Simmt, 2000). It is this part of Pirie and Kieren's view on mathematical understanding that emphasizes the notion that mathematical inter-activity as critically important for mathematical learning to grow. Moreover, Pirie and Kieren define mathematical understanding as the embodiment of all verbal, physical, and written acts:

> Mathematical understanding can be characterized as leveled but non-linear. It is a recursive phenomenon and recursion is seen to occur when thinking moves between levels of sophistication. Indeed each level of understanding is contained within succeeding levels. Any particular level is dependent on the forms and processes within and, further is constrained by those without. (Pirie & Kieren, 1989, p. 8)

A Descriptive Not Prescriptive Model

Pirie and Kieren make it clear that their model is not intended to be used to define or prescribe a particular sequence of static levels which constitute students' mathematical learning but rather, a way of conceptualizing the learning of mathematics as unpredictable and complex phenomena. As well, Pirie and Kieren (1994b) do not distinguish mathematical growth as being monological pathways, or privilege one's fluency to use formal language and mathematical symbols as representing formal mathematical understandings. Mathematical understanding is not only assumed to grow in complexity through the learner or collective unity's outward movement, but also from inward movement or, what they call, folding back to previous levels of knowing. Folding back is not a redoing of what has already been done, but moves the learner or group of learners back to inner levels of mathematical knowings where they will reintegrate understandings as a result of the perturbations experienced in previous outer levels before moving on (Kieren & Pirie, 1991; Towers, Martin, & Pirie, 2000). This model also reflects the notion of mathematical knowings existing simultaneously as a product, producer, and process (B. Davis, 1996; Kieren, Simmt, Gordon Calvert, & Reid, 1996). In this way, Pirie and Kieren advocate for learning settings that encourage students' engagement in folding back and enable their self-referencing, remembering, and reintegration of mathematical knowings to occur.

Thus, the Pirie-Kieren model of mathematical growth reflects an enactive perspective because it focuses specifically on the complex, co-emergent, and unpredictable nature of mathematical understanding. Mathematical understanding is conceived as occurring through interrelated, fluid processes and evolving in a fractal-like manner. They describe the model's structure as neither hierarchical nor linear. The realms of mathematical knowings found within this model exist as embedded, unbounded circles that are self-similar and compatible with one another. Moreover, the Pirie-Kieren model reflects Maturana and Varela's (1987) conception that "all doing is knowing, and all knowing is doing" (p. 26) because it locates primitive doing or knowing as being the roots from which all other mathematical knowings emerge (see Figure 23).

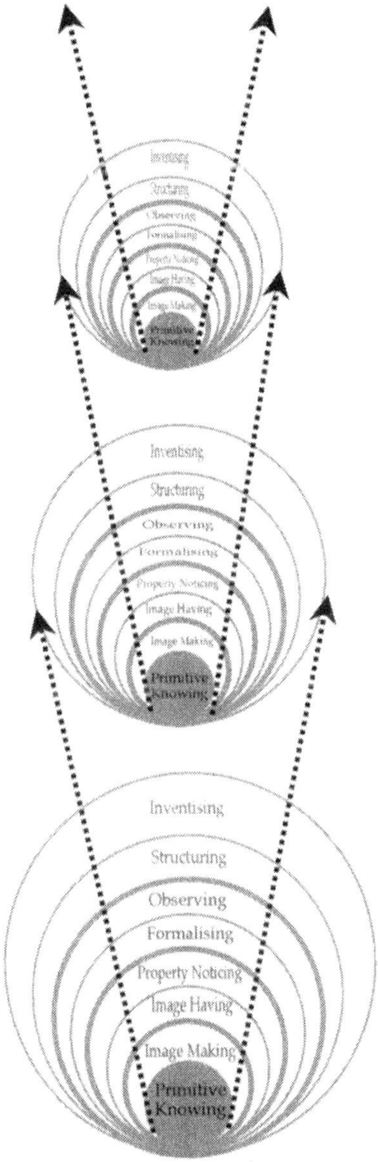

Figure 23. Model illustrating primitive knowing as the source of all other mathematical knowledge (adapted from, Kieren, 1990).

The Improvisational Nature of Collective Mathematical Understanding

Further work conducted by Kieren and Simmt (2002) and more recently, by Martin and Towers (2009, 2010) focuses on the use of the Pirie-Kieren theory to observe the growth of collective mathematical understanding. Although Pirie and Kieren's theory has the potential to be an interpretive lens through which to describe both individual and collective understanding, earlier studies focused primarily on personal mathematical growth and did not explicate how the theory might be used in making sense of the mathematical understanding that evolves from collectives. These studies share important assumptions about collective understanding with other bodies of research in mathematics education. For example, in keeping with Bowers and Nickerson (2002) as well as Cobb (1999), is the previously mentioned view embedded in the Pirie-Kieren theory that takes collective understanding to be an emergent phenomenon evolving within social-cultural contexts. There is also general agreement across the different bodies of research that while individual and collective understandings are both integral to mathematical understanding, each has qualities that distinguish one from the other. For instance, while personal understanding exists as mathematical ideas or actions that are observable characteristics of an individual, collective understanding happens at a different level where one member's idea or action is taken up by other members of the group and through reworking and elaboration, it becomes part of the group's way of making sense of the mathematics at hand.

This view is consistent with perspectives drawing from complexity science (e.g., Davis and Simmt, 2003) and emphasizes the notion that collective understanding is neither that which is achieved by one member of a group nor that which occurs by fitting together the individual group members' understandings. Instead, collective understanding is the mathematical coherence that results from the spontaneous, moment-to-moment collaboration of a group or members of it as a unified whole. This kind of understanding contrasts with that which is considered to be *taken as shared* because rather than being a matter of members establishing mathematical compatibility and norms for engaging in group work, it is about the understanding that arises out of and becomes embodied in the group's mathematical activity.

By bringing together the Pirie Kieren Theory and improvisational theory, Martin, Towers, and Pirie (2006) not only maintain collective understanding as dynamical growth locatable within the eight levels of the Pirie Kieren model (see Figure 21) but also provide insight into detailing the mathematical activity that occurs within collectives. Collective mathematical activity has been generally characterized as a complex, creative, and emergent process. Martin, Towers, and Pirie elaborate on how these qualities are enacted by revealing the improvisational ways in which groups of learners work collectively on mathematical tasks. The authors conceive collective actions as being "dependent on and contingent upon the actions of the others in the group" (2006, p. 156) and for this reason use the term, *coaction*.

They identify five aspects of coaction: *potential for pathways, striking a groove, etiquette, importance of the group mind,* and *pursuing a better idea.* Potential for pathways is a characteristic of coaction that refers to the emergence of many ideas and ways of thinking about the mathematics at hand. Striking a groove is a term used by Berliner (1994) that refers to the moment at which a group of musicians coalesce and play with the same sense of rhythm. In doing so, the musical performance takes on a life of its own. Martin, Towers and Pirie apply this term to the context of the mathematics classroom and make the point that striking a groove while working collectively on a mathematical task is not about negotiating a set of rules or establishing norms about how a group works together but rather, it refers to the fluid, responsive, and ongoing coordination of the members' mathematical activity. And similar to the notion of etiquette in a musical context where each musician attends to what is being played and responds accordingly, mathematical learners engaged in coaction, means that they are paying attention and responding to the mathematics that is emerging from the group's work, changing tact if necessary. Still further, when a group of learners do mathematical work in a shared or mutual way, their activity possesses the importance of group mind; this differs from a group of learners working individually on the same task or working in a manner that is based solely on one person's ideas. Lastly, Martin, Towers, and Pirie describe the pursuit of a better idea as occurring when other personal goals are abandoned in order for the group to take up an idea that is collectively deemed to be the most viable for solving a particular mathematical task.

The authors' research compliments and extends previous and current theoretical work being conducted in mathematics education. By continuing to portray mathematical understanding as growth, Martin, Towers, and Pirie offer a way to view up close, the moment-to-moment emergence of mathematical understanding as well as the creative processes by which such understanding evolves.

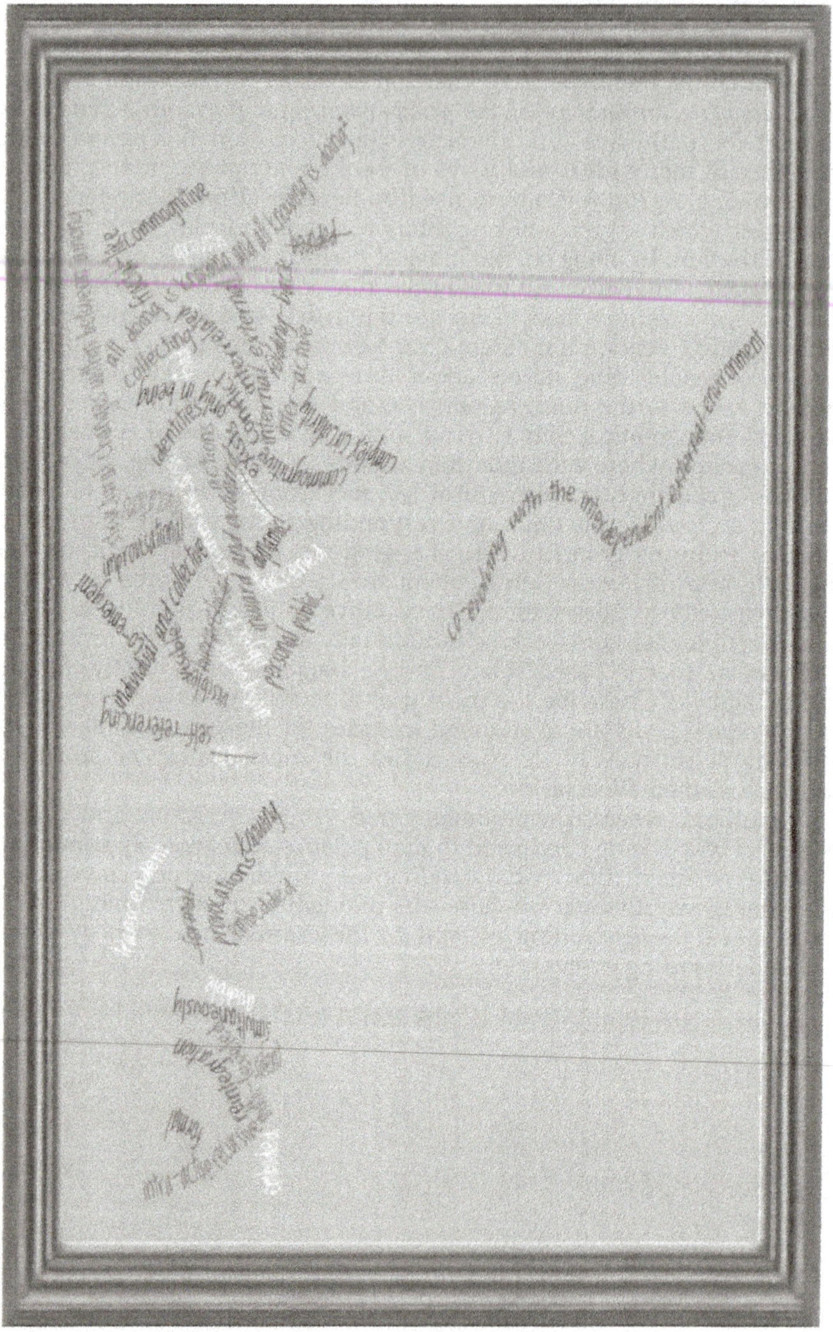
co-evolving with the interdependent external environment
self-referencing
co-emergent
individual and collective
collecting
personal public
simultaneously

Despite the diversity of interpretations taken by mathematics educators regarding the ways in which mathematical understanding is portrayed and the manners in which it is conceived to develop, there does seem to be general agreement across the bodies of research that "good" mathematical understanding involves the integration of informal and formal mathematical knowledge, that it is flexible and that it can be applied to respond to many different situations.

The theoretical portraits located in constructivism emphasize the building of mathematical knowledge as schemas whereby learning involves the progression through specific stages as well as the maneuvering of one's mathematics over a variety of conceptual obstacles. The perspectives taken by the authors in the second collection of literature differ slightly from those seen in the first as these researchers seek to interpret the dynamic nature of mathematical understanding and capture the associated forms of knowledge as interrelated phenomena. And finally, in the third grouping of research, the work of mathematics educators who share an enactive conception was examined. These authors in a more ecological manner, highlight the co-emergent and complex nature of mathematical understanding, how it is individually and collectively brought into being, and the embodied forms in which it exists.

Having taken a good look at the metaphorical furnishings that Jennifer had arranged comfortably around her teaching and revealed the diversity in how mathematics[1] and mathematical understanding[2] are portrayed, she now faced the task of deciding whether (and why) these furnishings suited the ecological mind-space in which she dwelled. Jennifer needed to consider the kinds of thinking that her metaphorical manners enabled and, if they were not engendering ecologically coherent ways in which to conceptualize her mathematics teaching, what furnishings might.

New Furniture

So what of my metaphor of mathematics curricula as jigsaw puzzles? The metaphor creates the image of a mathematics curriculum as being a set of pieces — like concepts, skills, and domains of mathematics. Complicated maybe, but as such, are doable and undoable pieces and as a whole, clearly defined and separately visible. So, when given to students and the pieces are assembled correctly, they reveal a coherent picture from its interlocking parts. Okay yes, the metaphor is a very tidy and systematic way for a teacher to think about mathematics curricula. But what this metaphor does not do is reveal the ecological qualities that distinguish teaching and learning of mathematics as dynamic and complex.

B. Davis, Sumara, and Luce-Kapler make the distinction between complicated and complex thinking when they describe complicated thinking as that which aims to reduce phenomena to elemental components, root causes, and fundamental laws.[3] *They use the example of a clock as a complicated mechanism and state that a complicated understanding of it would involve detailed knowledge of each of its parts*[4] *and how the clock could be disassembled and reassembled. In contrast, the authors explain that a complex comprehension of a clock entails not only an understanding of its parts and how they work together but also, the role that is played by the clock is necessarily embedded in [and thus affects as it is affected by] social and natural environments.*[5]

The authors make the important point that something conceptualized in a complicated manner as in the case of the clock, C = A + B *which implies that* C *(i.e., the clock) can be taken apart and put back together again. Thinking in a complex way however, assumes that* C *depends on* A *and* B *but that at the same time, it exists as something other than just* A + B. *Take for example, a cake, which can be considered to be a complex entity. Once the ingredients have been mixed together and baked, you cannot take it apart again to get back what went into making it. The cake exists as a complex form because of the reaction between the ingredients and its environment.*

So even if a teacher managed to design a mathematics program that fit mathematical topics together in a way that produced a complete picture, the curriculum would remain static. It would still be a set of distinct pieces and hence, a complicated not complex curricular form. The program would only be a product of its parts and not something that possesses the potential for possibilities greater than that or a curriculum that embodies awareness for the role it plays in the whole of teaching and learning mathematics. What I need is a complex view of the curriculum — a view that is more than a complicated one. I need a new metaphor.

(Weeks later…) *An ecologically coherent metaphor of a mathematics curriculum must be one that creates an image of a flexible, dynamic network that co-evolves as a result of the interactions of the teacher, students, and the environment. A mathematics curriculum then, might not be thought of as a commoditized thing and that which prescribes what teachers are to teach and what students are to learn.*

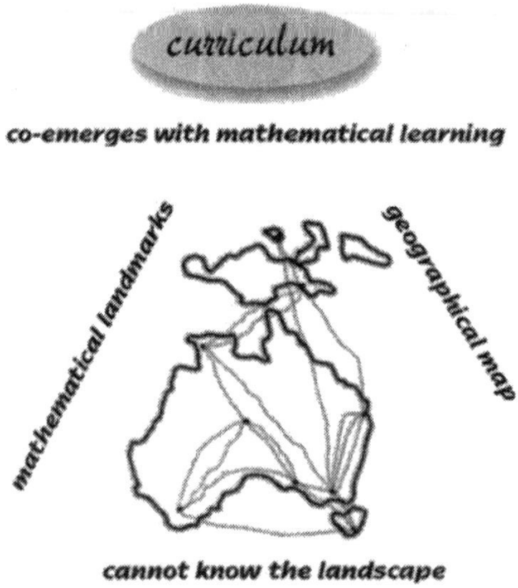

G. Bateson's elaboration of Korzybski's map metaphor is effective in that mathematics curricula can be understood in light of an ecological perspective. Envisioned as a map, a teacher may locate topics, concepts, and skills as important landmarks for the class' learning. Once these locations are marked out, the teacher can then think about how they are connected to one another. Inherent in this is the idea that what cannot be sketched out in advance are the actual paths on which the students will travel to get to the mathematical locations or the understandings they will establish when and after they come to these sites. So although a teacher might mark out mathematical landmarks, it is impossible to predict the conditions of the ever-changing landscape — the actual terrain that the class encounters when engaged in the mathematical studies.

In light of this, students' mathematical understanding can be likened to what Varela, Thompson, and Rosch describe as "paths that are laid down in walking."[6] And like paths, what the students come to learn is rarely if ever straight forward but

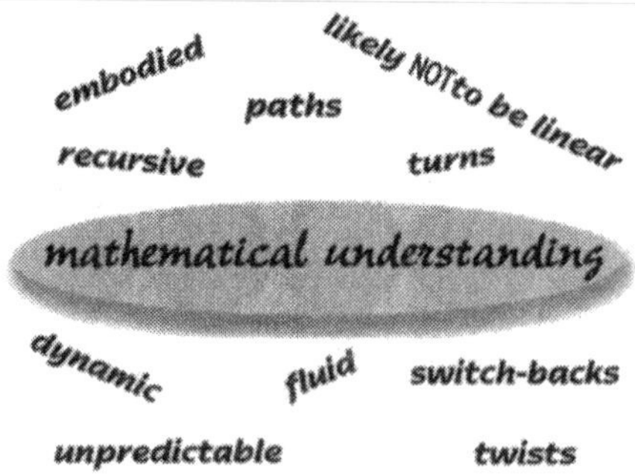

instead spreads out in several directions and entails twists, turns and switchbacks. Mathematics curricula, understanding, and learning imagined in this manner — as events which are lived — distinguish these as co-emergent phenomena that are brought into being through the students, their teacher, and mathematical settings.

Differently from before, I have come to envision the mathematics class as an image of teacher and learners interacting within individual and collective mathematical realms as nested systems, but not discrete ones. And, the larger human and nonhuman environment is not an external entity but an interdependent one. Once a moot point in my thinking, it is now a critical one. Just as the teacher and students' mathematical knowings impact and shape the classroom environment, the environment now changed, impacts and shapes the understandings that will take place for the class and so on. These realms exist as co-evolving systems —dynamic and recursively related to one another.

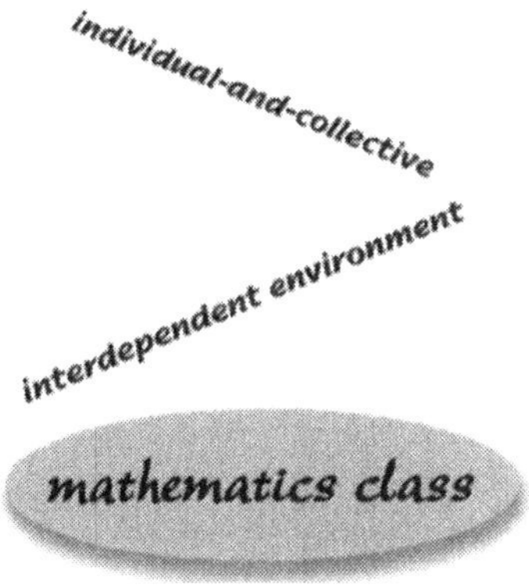

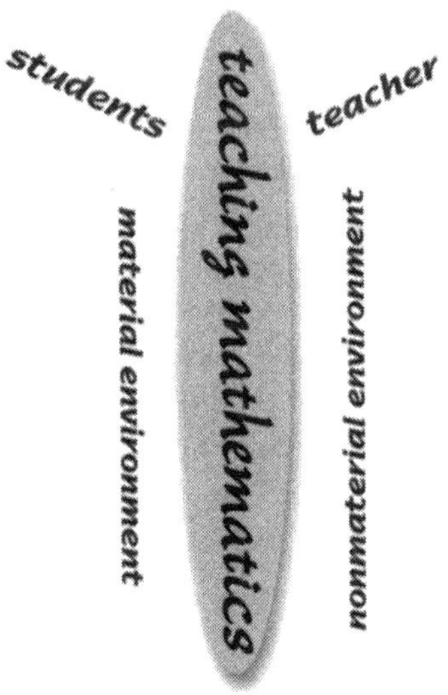

Teaching encompasses more than me as the teacher; it includes the students as well as the material and nonmaterial environment. Of course, I am a source of provocation and invocation of students' mathematical learning but I am not the only one. As oikos,[7] teaching and learning exist within systemic relationships of difference. The class perceived as a system of individual and collective relationships within the material and nonmaterial environment, enables learning to be viewed as that which is occasioned by anyone or anything.

When I think about a metaphor for mathematics, instead of notions of static integration or separate strands of algebra, geometry and so on, I imagine them to be fluid and co-emergent entities that make mathematics a dynamic living whole. In this way, mathematics can be conceived as a shape-shifter[8] *— arising and metamorphosing as it interacts with the contexts of which it is a part. Take for example, a linear equation as one form of mathematics represented algebraically and the equation as a graph as being another... or a pattern as a numerical sequence or as a three dimensional structure.*

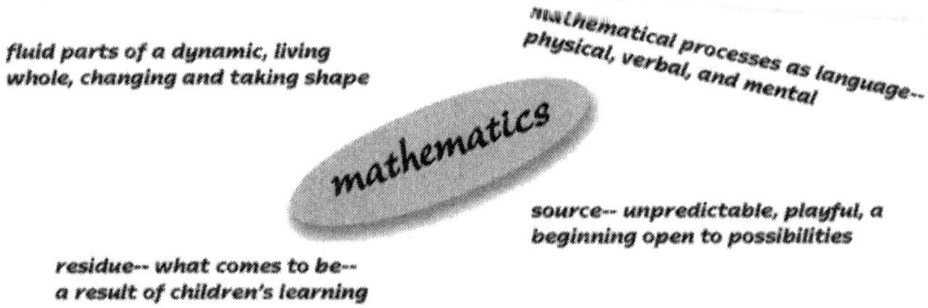

Now imaginable are other equally critical differences. Given this mind-space, mathematics can emerge as a residue or a source of students' learning. It does not have to be kept as a product that is produced within the sequential synchronicity of predetermined outputs fed by particular inputs. It is not framed in a deterministic or a predeterministic view of classroom mathematics. Here, R. Davis'[9] *notion of mathematical residue serves as a useful beginning point for re-imagining mathematics in the classroom. As residue, it is assumed that neither the origins nor the processes by which a particular mathematics arises can be precisely located. It arises from the mathematical languaging of the individual and larger collective(s). Mathematical processes such as problem solving, connecting, reasoning, and expressing can be understood as mathematical language; that is, they are the physical, mental, verbal, spatial and temporal manners in which we think, (inter)act, and exist mathematically. And, mathematics as a source for students' learning embeds a sense of unpredictability and playfulness — a beginning that is open to all kinds of possibilities. In a radically different metaphorical manner, focus is not only on the mathematics at hand, not only on mathematics as individual and collective knowing, but also on mathematics as it resides seamlessly and all at once within past, present and future contexts of students' learning.*

EDUCATION

"JUST HAND THEM DOWN THE MATHEMATICS" ... OR NOT?!

WHEN MATHEMATICS is imagined and enacted as objectified, static knowledge that is to be traditionally passed down from one generation to the next, the teaching and learning of mathematics is disabled from ever becoming anything else. Under the air of hand-me-downs, it is easy to understand why mathematics is taught and learned out of a sense of obligation or contempt rather than a sense of open desire or wonder. It is also understandable why mathematics is all too often considered as that which is to be mastered rather than that which is to be understood. In commoditizing mathematics, we make absurd, the possibility for us as teachers and our students to perceive mathematics as anything else but a fixed entity. In this way of conceiving mathematics, we make it inconceivable for school mathematics to become something other than just a collection of hand-me-downs.

The embeddedness of these images within one's taken for granted ways of thinking about mathematics not only makes it natural for us to assume mathematics to be an inanimate thing, but in doing so, locates mathematics as that which exists — out there. Given this mindset, it is not surprising that a teacher would feel impelled to set the class on a straight and narrow, one-way course to become collectors of mathematics. Given this mindset, it makes sense to ingrain the ritualistic practice of acquiring mathematics into school mathematics unit and lesson plans, methods of assessment, and enacted in the classroom; product oriented practices that focus on desired, expected, or even measurable outcomes of instruction — that after instruction — the students will have mastered the mathematics taught in the lesson before moving on to the next part of the curriculum. Of course, the ways in which students are instructed to take possession of their mathematical hand-me-downs of concepts, skills, and even attitudes may vary. Still, teaching by telling, engaging students in hunting for, having them seek out hidden mathematics in real world contexts, and even explorations de-

(continued on page 84)

(continued from page 8) signed for students' discovery of mathematics all keep alive, the tradition of handing down mathematics.

Moreover, when product-oriented ways of thinking about school mathematics are coupled with a back to basics mentality, the teaching and learning of mathematics becomes subjected to the weigh scale of how much in regard to the amount of mathematical facts and skills that students are to learn and little or no emphasis is placed on such things as their mathematical thinking or conceptual understanding. Given this mindset, mathematical processes such as those identified by the National Council of Teachers of Mathematics[1] as: problem solving, reasoning, communicating, connecting, and representing, would likely be deemed by most teachers as not essential. Viewed as additional knowledge,[2] teaching that attends to the development of mathematical processses would then depend on whether students have acquired first, the prespecified mathematical facts and skills with which to process the mathematics.

The point here is that when students are taught to learn mathematics in the tradition of hand-me-downs and as a product oriented matter of collecting, hunting down, or retrieving pieces of knowledge, this creates the impossibility for mathematics to be taught and learned in ways that enable it to arise as living and animate.

Now, identifying the limitations of how mathematics exists in the classroom and the possibility of it becoming something else is all fine and good. But in doing so, means that the conversation does not end here. Rather, it opens up a whole host of questions that require further interrogation such as:

- How might an ecological way of thinking help to reconsider such taken for granted perceptions and re-imagine a more responsive view for the teaching and learning of classroom mathematics?
- What shifts in thinking are necessary to conceive classroom mathematics as something other than a line of hand-me-downs from teacher to student?
- **What would it mean if we assumed mathematics to be embodied?**
- **How might problem solving, reasoning, communicating, connecting, and expressing be understood as integral to the learning of mathematics and not as additional knowledge?**

If we take seriously the view that mathematics is an ever-changing entity perceived, created, and embodied as we interact with the world, then it doesn't make sense for mathematics to be conceived as a fixed, inanimate, disconnected thing that exists — out there. When we assume that the only mathematics we know or can ever know emerges from our patterns of living as biological and social-cultural beings, then mathematics is not an objective, universal, pre-existent reality but that which is alive and necessarily constitutive in nature. This means that mathematics takes place in the praxis of living in language[3] and its coherence is dependent on those who bring it into being.

Languaging Emergent Mathematics

Scripting an Unscripted Lesson

The children gathered on the carpet to listen to me read the story, *Even Steven and Odd Todd.* The beginning of this lesson was simple – there was no formal introduction, no preamble, not even a motivational hook. I just opened the book and read. This was not because I had come unprepared; in fact, I had. One of the reasons for why I chose to read this particular story was that it had both constraining and liberating qualities. What I mean by this is that while the story certainly put constraints on what the class focused their attention and activity on – even and odd numbers, at the same time, the story was also liberating in that it did not tell the children what or how to think but provided a provocative and open space in which the potential for a rich range of conceptual possibilities could be occasioned by the children's sense-making of numbers. Keeping this in mind while I planned this lesson, I knew that I did not want to tell the children to *look for*, *listen for*, or *think about* anything specifically. I did not want to preface the book by telling them that it was a humorous tale about two cousins who are different from each other in one particular way. I did not wish to tell them that in my earlier reading of the book I had found out that for Even Steven, life unfolds in a pattern of even numbers and for Odd Todd, life is all about odd numbers. It was not that I wanted to keep any of the mathematics from the children but rather, I set out to create a teaching and learning space where the class experienced the story as it happened, in that particular moment. I wanted the mathematics to emerge in a different manner than a predetermined, predictable one. In not creating a pre-scripted lesson that was about me, the teacher, identifying the mathematics that the children were to find in the story, this lesson remained open to and contingent on the mathematical ideas that we would bring forth as we listened and responded to the book as a class.

A moment of digressive thought:

Problem Solving
Reasoning and Proof
Communication
Connections
Representation (or, Expressing)

If one glances at the NCTM's *list of the five process standards or reads through the descriptions that accompany them, it is possible for teachers to interpret them as five discrete mathematical skills. Given this, it is understandable why a teacher might then present them to the class as five separate topics.*

It is only when carefully reading through the NCTM's *document, The Principles and Standards for School Mathematics and making note of the Council's statement that, processes can be learned within the Content Standards, and content can be learned within the Process Standards,*[4] *that a nonlinear image of mathematics and mathematical processes is possible. In contrast to linear, mechanistic forms of thinking that would have us imagine mathematical processes to be mechanisms or devices that act as a conduit through which we transmit mathematics content into students, the* NCTM *proposes that mathematics content and mathematical processes be understood as reciprocal in nature. The Council takes the position that the process standards are not to be conceived as additional knowledge and certainly not as a means to a linear end in mathematics classrooms. However, because this point is communicated briefly in the document, it does not make prominent, this nonlinear and co-emergent image of mathematics and mathematical processes. Consequently, it remains faint and is easily overlooked by the reader who reads from a background of traditionally mechanistic ways of thinking.*

Differently, if we set the image within an ecological realm, it immediately conjures up notions such as complexity, circularity, and recursion, all of which help to bring forth an implicit understanding that:

students' mathematics

arise
can from and be
shaped by
mathematical processes,

mathematical processes
can
also arise
from and be
shaped
by

students' mathematics.

Moving deeper into the languaging space of ecology, what is brought forth is mathematics content and mathematical processes as enveloping and co-emergent entities. Such thinking not only enables an awareness for the complex circularity that exists with-and-in content and processes but also, a mindfulness for the interdependence that defines them as being inseparably part of each other. And it is not that they simply exist in a cyclical sense; that one prompts the formation of the other, but in an integral manner, mathematics content and mathematical processes emerge and evolve together in relation with each other.

Now if we take the NCTM's *process standards and consider them from an ecological space that includes Maturana's definition of languaging and Sfard's notion of mathematical thinking as communication,*[5] *the processes identified as: problem solving, communicating, reasoning, connecting, and expressing can be understood as forms of mathematical language. As explained earlier, Maturana's description of languaging entails how collective unities evolve through their physical, verbal, and mental linguistic manners of thinking, acting, and existing. Mathematical processes as forms of mathematical language are the mathematical patterns of thinking, interacting, and being that enable a class, for example, to exist and develop as a collective entity. If we conceive mathematical language and mathematics to co-exist and seamlessly co-evolve with each other, then each is fundamental to the other and thus, in how we teach mathematics and how students learn mathematics.*

When classroom mathematics is envisioned as a living entity that emerges with the class' mathematical and social-cultural forms of languaging, it becomes something more than just an end in itself. Mathematics becomes that which results from students' learning as well as that which serves as a beginning for their further learning. Thus, mathematics exists as both residue and source.[6] When mathematics arises from students' mental intra-actions or social interactions, it can be conceived as a residue or, an outcome of their learning. This contrasts with pedagogical views that assume learning can be prespecified as learning outcomes. An ecological examination does not focus on what the child should know but rather, on what mathematics the child actually comes to know. For instance, if a child explores how different sets of objects can be arranged in smaller, equal groups and arrives at an understanding that division makes smaller, this would be considered to be a mathematical residue of that child's learning.

Mathematical Language, Languaging, & Residues of Learning

(A continuation of Scripting an Unscripted Lesson)

As I continued to read the story to the class, the children began calling out differences that they noticed between the two characters:

"Even Steven has six cats, eight gerbils, and ten goldfish."

"... and twelve sprinklers in his garden!"

"Even Steven gets up every morning at eight o'clock sharp."

" Odd Todd likes to get up at nine o'clock sharp."

"Odd Todd rides a tricycle."

"Even Steven has four bicycles — they have two wheels each."

"Odd Todd has five buttons on his jacket and Even Steven has six on his shirt."

While I recorded the children's observations onto a large piece of chart paper Danica looked at what was being written down, glanced away for a moment, and then announced, "Even Steven only likes things that are two, four, six, eight, ten, twelve, and so on... and Odd Todd only likes things that are one, three, five, seven, nine, and eleven."

The whole class nodded and smiled in agreement.

Mark then added, "Even Steven likes EVEN numbers."

This was immediately followed by Robby's comment, "and Odd Todd only likes ODD numbers."

This time however, only some of the class nodded or responded with "yeah!" while other children said nothing, looked puzzled, or exclaimed, "what?!"

"Numbers that end in zero, two, four, six, and eight are even numbers and numbers that end with one, three, five, seven, and nine are called odd numbers," stated Mark.

And even though Robby was nodding enthusastically during Mark's explanation, the class still reacted with a mix of nods, furrowed brows, and a bunch of "what?!s."

I jumped into the conversation and agreed with the two boys, "Yes, that's one way of thinking about numbers as being of two different kinds." I then began to push this space that Mark and Robby had opened a little further. Focusing on the two sets of numbers, I posed a question to the class: "If two, four, six, eight," I said, "and numbers that end with zero, two, four, six, or eight, such as ten that is written as one-***zero***, twelve that is written as one-***two***... thirty-***four***... sixty-***eight***, can be described as being *even*," I said while recording 2, 4, 6, 8, 10, 12, 34, 68 on a piece of chart paper, "and one, three, five, seven, and nine, as well as numbers that end with one, three, five, seven, or nine such as eighty-***one***, forty-***three***, eighty-***five***, sixty-***seven***, twenty-***nine*** can be described as being *odd*," I added as I jotted down 1, 3, 5, 7, 9, 81, 43, 85, 67, 29 on another piece of chart paper, "*what* is it that makes certain numbers *even* and other numbers *odd*?!... besides just looking at the digit that they end with?!"

Silence. The children said nothing. One second... two seconds... three seconds... waiting. They simply stared back at me, shrugging their shoulders. Michelle leapt into the conversation and conjectured that "*even* numbers – you add two to them... two, four, six, eight, ten."

Moving into her space of thinking, I poked around a bit. "Yes, this is true... but what about one, three, five, seven... aren't you also adding two to one to get three and two to five to get seven...?" I asked as I pointed to the series of numbers that I had written on the chart paper for "odd numbers."

"Oh yeah," Michelle said, smiling.

"As well," I added, "think about Even Steven... did he do everything in a sequence of two then four, then six, then eight...? or Odd Todd, did he do everything in a sequence of one, three, five, seven, and so on?"

"Hmmm... no," Michelle replied, shaking her head from side to side.

The conversation came to another standstill. I asked the class to form a large circle on the carpet. As I moved around the outside of the circle, I randomly asked children to build one of the odd or even numbers listed on the chart paper with counters and place it in the middle of the circle. By doing so I told them, we could continue the exploration but this time, take a closer look at the numbers in their physical form. Although the children did not arrange the numbers they had built in any particular order except for separating what they assumed to be even numbers from what they took to be odd numbers, we soon had physical expressions of 2, 4, 6, 8, 10, 12 and of 1, 3, 5, 7, and 9. I asked the class to look at the 'built' numbers, talk with one another, and see if there was anything that made the two sets of numbers different from each other; in other words, was there anything that was *even* about the even numbers or *odd* about the odd numbers.

After working with the students on either side of her, Shelby raised her hand to speak. "The four, it's even because it has two and two," pointing to the group of four's two pairs of two counters.

Some of the other students and I nodded and smiled.

"Yes. That makes sense... we can think of four as being evenly two and two," I said, replying to Shelby's comment.

Taking the idea that the group seemed to be embracing as making some kind of sense to them, I recorded it on the chart paper with their other ideas and encouraged the class to pick this idea up and play with it some more. "Okay, what about the other numbers that we've built here? Can we take Shelby's idea and... is there anything that we could say about the other numbers, using her idea of even twos?"

Without any talk, the class broke into smaller working groups. Reaching into the middle of the circle, the children arranged the counters for each of the even numbers by twos. When they had all finished working, I asked them what sense, if any, might they make from what they had done.

They explained to me that two was an "even two", four was "two even twos", six was "three even twos" and so on. And when I posed questions about larger numbers like "what about thirty-six?," I quickly got responses like, "that'd be eighteen even twos." I recorded these ideas onto the chart paper about even numbers.

Noticing that the children had not said or done anything with the set of odd numbers they had built, I pointed to them and asked, "what about these?"

This prompted the children who were sitting nearest the odd numbers to reach into the middle of the circle and begin to move the counters about. Soon comments such as "this one doesn't have a partner," "neither does this one," and "none of them do," surfaced in the conversation as the students arranged each of the numbers by twos.

Then, just as I had done with the even numbers, I asked the class what they could say about larger numbers, "what about twenty-nine?"

Here, I got replies of, "it would have fourteen even twos and one leftover" and, "if you put it into partners, one would always be left lonely."

I added these responses to the chart paper about odd numbers.

As we sat back and looked at the children's recorded ideas about even and odd numbers and the two sets of numbers now arranged evenly or unevenly by twos, I asked the class if there was any sense in looking at these numbers as being two different kinds and what sense, if any, was there in Mark's claim that you simply look at the last digit of a number to determine whether it is even or odd. Some children reiterated that by building or visualizing a number as a set of counters and then "partnering up" the counters, a number was odd if there was a remainder of one and it was even if it had no remainder. Other children concluded that if a number can "be split into equal halves" it is even and if it cannot, then it is odd. Still other children replied that if you arrive at the particular number by counting on by twos from zero, the number is even but if you count *past* the number, the number is odd. These two other strategies were recorded onto the chart papers.

Returning to Mark's claim and elaborating on it, Danica stated that, "A number bigger than ten, like forty-eight, the forty is groups of ten... even... and the eight is four even twos... it's an even number."

"And forty-nine?" I asked.

"The forty is four groups of ten... that's even, and nine is an odd number because it has four even twos and one leftover, so it's an odd number," explained Danica.

"Are you saying that you don't need to look at the other digits, only the last one to determine if the number is odd or even?" I asked.

"Yes... because they are always even... tens... hundreds... they are always something-zero," she said.

Danica then directed me as I recorded these ideas as diagrams; using 48 for the poster about even numbers and 49 for the one about odd numbers. The diagrams illustrated that when these numbers are grouped into tens and ones and then into groups of two, it is the last digit of each number that determines if the number is even or odd; that is, whether the last digit as single units can be put into groups of two.

With the children's strategies scattered on the two posters, I asked the class to reflect on what was the same or different about them. The children's responses led them to conclude that all of the methods involved the idea of even twos but that they differed in "how they were two." For instance, while the class considered the strategies of partners, even twos, and counting by twos from zero to be identical because all three involved imagining a number as being composed of groups of two, the strategy of "equal halves" however, determined whether the number could be split even*ly into* two *parts.* Moreover and similar to Danica's explanation and diagrams, the class related the idea of even twos as being inherent in the place value method of looking at the last digit in two ways. Firstly, because any number greater than nine as a numeral has digits to the left of its ones' column which represent groups of tens or hundreds or thousands and so on, these values themselves always appear as "something-zero" numerals; therefore, because the children already knew that any number ending in zero can be arranged evenly into groups of two, such parts of the original number are assumed to be even. Secondly, given this, only the digit in the numeral's ones' column needs to be assessed as to whether it can be put into groups of two.

Reflecting on this lesson, I am reminded of Theodore Roszak's book, The Cult of Information. In it, he explains that as humans, it is ideas and not information that we think with; through the recursive integration of patterns which emerge from our lived experience, it is ideas that form and from which information arises. And so too, the class integrated the ideas of their lived mathematical experiences to form meaningful information about odd and even numbers. They did not blindly accept Mark's mathematical fact that even numbers are numbers that end in 0, 2, 4, 6, or 8 and odd numbers are ones that end in 1, 3, 5, 7, or 9 or, Shelby's conjecture of even twos as truths. Rather, the class jumped into these spaces to explore them further, evaluating whether the ideas made sense conceptually by reflecting on what they already knew about numbers as well as coming up with and assessing their own conjectures in order to develop individual and collective meaning for odd and even numbers.

Another afterthought: A curious space that was not opened because neither the students nor I thought of it at the time (and the story frames the reader's thinking to assume that all numbers are even or odd) was this:

Is there ever a number that is NOT odd or even?

On another occasion this might have arisen naturally and taken the class to a different mathematical place. Nevertheless, this lesson served as an example of how the mathematics of odd and even numbers was brought forth by the class' mathematical languaging of the story through their physical building of numbers, mental images of numbers as a collection of objects and symbols, as well as powerful metaphors of even twos, partners, lonely ones and equal halves.

The mathematical residue of the students' learning was unpredictable and distinct yet integrated. Just as it is possible to have many different shades of a certain colour, the class demonstrated that there are many different ways to make sense of even and odd numbers. Embedded in these ways of making sense were understandings that varied from each other in at least one way as being rote or conceptual in nature or in the contextual or mathematical manner in which the understanding was grounded. At the same time and again, similar to the different shades of a particular colour, each way of thinking blended with the others through the notion of two, creating a sense of depth and dimension to the students' eventual personal and collective understandings of what it meant for a number to be even or odd.[7]

Differently from mathematical residue, when mathematics occasions students' curiosities, it exists as the source or the place where mathematical growth occurs. For example, posing the question — is it true that division always makes things smaller? — enables a source from which many different mathematical directions and spaces for students to take their learning can occur. Mathematics as source makes room for students to move mathematically and to deepen their understandings. Given this, students can explore and establish ways of knowing for when and why such a meaning of division might be appropriate and also experience situations in which an understanding of when and why it may not make sense and mark out another place of knowing.

Arisings

(A Continuation of Mathematical Language, Languaging, & Residues of Learning)

Settling down in front of me again, the children looked up, telling me that they were ready to finish the story. I read on from where we had left off. It was when we came to the part in the story where Even Steven sets aside six pancakes for his lunch and Odd Todd comes along and eats three pancakes that the children got very excited. They stopped gazing at me and turned towards one another.

"He just made Even Steven's pancakes odd!" Mark exclaimed.

Michelle giggled and then whispered to Danica, "Odd Todd is very clever!"

Even after we continued and then finished reading the story, the children's chatter about how the even number of pancakes had been "turned into" an odd number of pancakes did not diminish. When I asked them what it was that had them so intrigued about the pancake incident, a flurry of responses came at me:

"I want to see what odd numbers I can make even!"

"Can ALL even numbers be made into odd ones?"

"I wonder how Odd Todd would eat his cousin's eight pancakes?... so they'd be odd."

"Or Even Steven's ten pancakes?!"

"Or his thirty-six pancakes?!!"

"I want to find out some other ways Even Steven could eat his twelve pancakes."

"Does an odd number and another odd number always make an even number?"

Here within the same lesson, the ideas of odd and even numbers arose again, not as mathematical residue but as mathematical sources. Despite the conclusion of the story, new beginnings sprung from the children's questions, opening spaces in which they could explore and deepen their understanding of odd and even numbers. And for most of the two days that followed, the class worked to examine these questions.

Do the metaphors with which we describe beginnings and ends really matter?

I suppose it is possible, one could argue, that even in mechanistic patterns of thinking, mathematics can exist as both a beginning and an end in students' learning. For example, rather than being defined in terms of source and residue, the metaphors of input and output could be used. By doing so, one might wonder whether in fact this does not make both ways of thinking pretty much the same. And if it does not, what exactly is the difference?

Yes of course, mathematics can be conceived in terms of inputs and outputs of student's learning. However, there are critical differences between these mechanistic metaphors and the ecological ones of residue and source. You see, by its very nature, output as a metaphor evokes the idea of classroom mathematics as a product produced or re-produced by a chain of learning events. And, input, used as a metaphor for such events conjures up notions of learning mathematics as being identifiable, measurable, or even prescribed components deemed necessary to produce a certain output, result, or product. Together, these metaphors embed a sensibility that assumes specific events which produce particular outcomes to be distinguished as such and that both the outcome and its inputs are predictable; further still, that they might even be predetermined. This kind of thinking serves to maintain a confining and reductionistic view of classroom mathematics.

In a radically different manner, the ecological metaphors for mathematics as being a residue and a source for students' learning do not give rise to a predeterministic or even deterministic view of classroom mathematics. Rather, mathematics as residue evokes the idea of something that has come to be, what is left or what remains as a result of students' learning. It is not a way of thinking which presupposes that the origins of a mathematical residue can be precisely located or that the process(es) or languaging acts by which it comes to be can be identified, specified, or replicated. In the previous lesson, it is clear that the ideas and the ways in which these were brought forth and related (or not) to even numbers, odd numbers, and to the other ideas themselves were indeed a residue of the class' responses to the story. Their mathematics was not a product produced by predetermined learning outcomes but rather an outcome of their learning. In other words, the students' mathematics and their understandings were dynamic in nature; contingent on and occasioned by their interactions as individuals and collectives. These interactions served to invoke the class to re-curse on what they knew about numbers and in doing so, provoked new ideas with which they could think. Emerging from their knowledge of skip counting, repeated addition, division, and base ten came strategies of even twos, partners, equal halves, counting by twos and place value integrating groups of two; all of which enabled the students to make conceptual sense of what it means for numbers to be odd or even.

As metaphors — residue and source — focus attention on learners' mathematics at hand; that is, the emergence of their mathematics and their understandings of it. Mathematics as a source for students' learning contrasts with the mechanistic image of mathematics as input in that the notion of source brings with it a sense of beginning; a beginning that is not concerned with determining or predicting what should follow but instead, highlights the need to be mindful and open toward the possibilities that might come about — anticipated or not. As seen in the students' study of even and odd numbers, mathematics as a source for the students' learning emerged several times throughout the lesson and took on many different forms; for example, at the start of a lesson as a rote definition of odd numbers, during the lesson as an image of four being two and two and as a series of problems posed by the class after the story had ended.

Imagining mathematics to be both residue and source infuses a sensibility that is open — as when the ground opens and water springs forth — and, ever-changing, as how one can never step into the same river twice. These two metaphors offer inherently different images and meanings with which to consider how mathematics might exist in the classroom; images and meanings that mechanistic ones cannot.

After moving deeper into that other mind-space,[8] Jennifer emerged with a new understanding of what it meant to teach mathematics. Her work in re-imagining metaphors that possessed an ecological sensibility was occasioned only because she now conceptualized her previous mechanistic mind-space and her current ecological one[9] as two different consensual domains. In this way, Jennifer recognized how each of the domains as languaging systems specified and reinforced a certain view of teaching classroom mathematics by highlighting and diminishing particular issues or concepts through very diverse metaphorical languages. For her, it was impossible to conceive language as simply a tool for communication.

We exist in language and it is through our being in language and languaging that we bring forth metaphors that invisibly yet powerfully become our ways of thinking, how we teach, our ways of researching, and ultimately, the kinds of places that are created in the mathematics classroom.

"Atmosphere," as the word suggests, is a vaporlike sphere which envelopes and affects everything.... The sense of mood or atmosphere is a profound part of our existence. By it we know the character of the world around us. Mood is a way of knowing and being in the world.... the way in which space is lived and experienced.

~ Max van Manen[1]

[E]very classroom, every school contains a certain atmosphere [sense of place]. The question is *not* whether there should be a pervasive atmosphere in the [classroom or the] school, but rather what kind is proper for it, worthy of it.

~ Max van Manen[2]

Sense of place. Place-making.

van Manen's thinking about atmosphere echoes the very notions being explored here of ecological thinking and of place.

Worthiness

The author moves the conversation into the realm of education and highlights the taken for grantedness of atmosphere. He identifies it as being that which profoundly shapes teaching and learning. van Manen makes important, the need to be aware of it, to consider the qualities that make up a particular atmosphere, and to exercise a mindfulness and care for the sense of place we bring forth for and together with our learners. This certainly implicates the examination of one's metaphors and the mathematics teaching and learning occasioned from the embodiment of them. In doing so, van Manen hopes to provoke us to question the worthiness of our actions in relation to the kind of place such rituals in teaching mathematics create.

Because this tacit knowledge influences both *what* and *how* learning occurs — for all participants.... In some instances, the teacher must keep up with the social reconstituting of taken-for-granted knowledge, and in other instances... the teacher should take a leadership role that can only be fulfilled by modeling and not simply by substituting a new set of taken-for-granted beliefs for the older ones.

~ Chet Bowers and David Flinders[3]

The leadership role that Bowers and Flinders speak of is important in developing one's teaching of mathematics. Relating with earlier sentiments, growth as a teacher is not about breaking away from one's entire teaching and replacing it with something else. Not only is the possibility of this questionable, such thinking only exacerbates 'this-or-that', 'either-or' attitudes and reactions. Differently, what Bowers and Flinders argue for is in keeping with what van Manen too seeks. Situated in the mathematics classroom, this entails a teacher's mindful consideration for the kind of oikos[4] being created for learners through what one chooses to conserve in one's ways of knowing and actions as a mathematics teacher.

For me, this has meant assessing the worthiness of the metaphors embedded in my teaching of mathematics, identifying which ones did not engender an ecological sensibility, and re-imagining ones that would. One might assume that doing so should make the enactment of these conceptual shifts natural and effortless given that I was now thinking in a theoretically different way. However, each day that I stepped into the classroom, I was confronted by the fact that:

AGAIN I am reminded that learning what it means to teach mathematics is not an automatic process. It is not smooth, it is not straightforward, and it certainly does not appear on demand.

even though I have consciously created differences in how I conceive mathematics teaching, in the classroom — as part of this place, I am still and always will be unconsciously embedded in a web of taken for granted relationships — taken for granted language — in which and by which I teach and students learn mathematics.[5]

The challenge then becomes, which relationships need to be a critical part of one's teaching consciousness?

•

•

•

How might one go about occasioning different spaces for teaching and learning?

•

Establish new relationships?

•

Enable an-other kind of place for mathematics to grow?

•

•

•

Assuming teaching to be praxis and not a practice that imparts the understanding that the teacher can never be separate from the learners themselves (and the metaphors they bring to their understanding of place), the events that take place, or the relationships that exist within the classroom. All co-emerge and co-evolve as living systems within systems dynamically in response to one another. One's teaching then, cannot be identified in terms of what it is so much but more so in terms of how it is — how one's teaching is an interdependent part of the whole. So for me, my work is concerned with how my mathematics teaching relates to the greater whole of embedding an ecological sense of place for mathematics in the classroom.

As well, I do not have any anxious temptations to systemATically take my new metaphors[6] and fit them into the classroom. By situating my study of mathematics teaching within the systemIc realm, the focus becomes grounded in understanding how my metaphors think me (in ecologically coherent ways) as I think with and in them. Once again moving off the mental line enables me to head towards that other space.

oikos...
relationships...
relationships as patterns...
... as patterns of difference.

*So in the same way that Maturana and Varela speak of knowledge as constitutive in nature; that is, as dynamic structural relationships within a living system and that which distinguishes one living system from another, G. Bateson's concept of mind as connecting patterns of differentiation is also what is at the heart of this work. Focusing on difference in an ecological manner means re-***COGNIZING** *difference as relationships and not as things.*

A difference that makes a difference in my teaching of mathematics as learning place is attending to the impact that my metaphorical patterns of thinking have on the oikos of it.

So what are the patterns of difference in my new metaphors?

paths that exist only as they are laid down in walking[7]

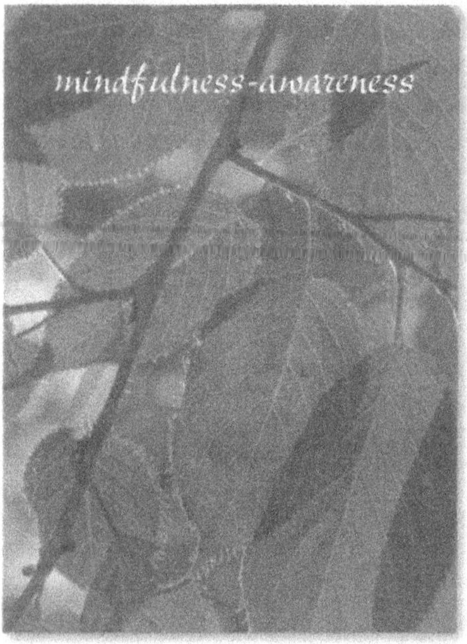

recursion

co-emergence

sustainability

diversity

individual-and-collective

possibilities

Model for the Growth of Mathematical Understanding[8]

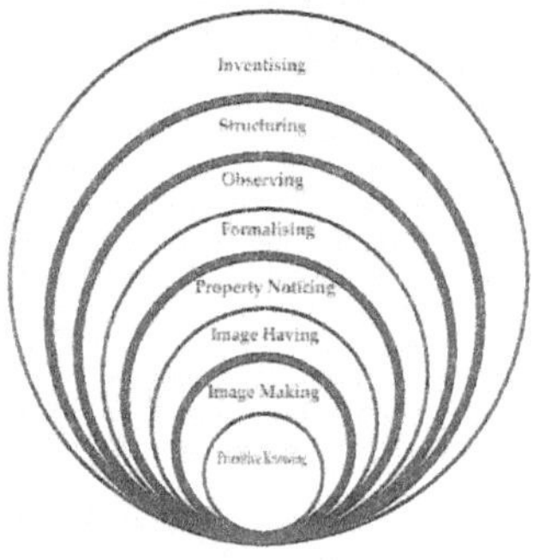

knowing is doing is being

~~linearity~~

flow

holistic

~~hierarchical~~

complex circularity

embodied cognition

interdependence

co-evolution

occasioning

~~predictability~~

In contrast to the mechanistic, linear, hierarchical, and predeterministic metaphors embedded in my earlier conceptualizations of teaching mathematics, these new metaphors of maps, paths, living systems, residue, and source evoke notions of growth, nonlinearity, networks, and unpredictability. These metaphors speak of mathematics teaching and learning as complex and recursive growth. They create the image of everything existing in fluid relationship to everything else. They enable the integration of mathematics to be understood not as a product of teaching and learning but as that which happens in the flow of the two. And, that neither mathematics teaching nor the learning of it nor mathematics itself for that matter, exist as fixed objects but all arise in the midst of dynamic interplay; varying in form and being occasioned in all kinds of manners.

These, I would say, are the patterns of difference in my metaphors.

What we conserve, what we wish to conserve in our living is what

Knowing the ground on which I walk.
Minding the paths I lay down in walking.

determines what can and what cannot change in our lives.[1]

Embedding and Rooting an Ecological Sense of Place for Mathematics in the Classroom

Children must feel that... the space, materials, and projects, values and sustains (sic) their interaction and communication.[1]

~ Carlina Rinaldi

Attending to the Physical Space in the Classroom

Stepping into a classroom is not unlike studying a painting. Both tell of relationships — the assumptions, ways of knowing, and the kind of experiences that have come to form a classroom's sense of place or the subject matter of a painting. In this manner, albeit through different methods, teachers and artists engage in the act of portraiture. In creating a painting, these relationships are expressed through the media and design of the work. The feeling of coldness and tension, for example, might be told through the painter's use of colour and texture whereas coldness and tension in the mathematics classroom is often conveyed through the austere and rigid manner in which the teacher makes the subject present. How a teacher structures the actual physical space for learning and the types of materials made available for students' use define the kinds of teaching-and-learning patterns that exist.

Stepping Into Jennifer's Classroom

What is obvious upon entering Jennifer's classroom is that even though there are no dividing walls or partitions, like an open concept home, there are many different areas that define the space. Each of these areas or what Jennifer considers to be physical spaces for learning are distinct in their purpose and the kinds of mathematical interactions they engender.

At their desks: Working on their own

At their desks: Working with a partner

Small group work at their desks

Large group work at the round table

The children's desks are arranged in 2 x 2 groupings in the middle of the room from which the chalkboard, chart board, and overhead projector screen can be seen. Although the desks store student belongings and each bares a child's name, the children frequently move about and use one another's desks. This area allows the students room to work on their own, with a partner, or in small groups. When in larger groups, the children often choose to work at the large round table located across from these desks.

Meeting place

On the floor

Two carpeted spaces can be found on either side of the desk area. The larger of the two is a meeting place of sorts, where Jennifer and her students gather in the morning and at the close of each school day. This is also the place where the class meets to share and examine the mathematics they have been working on. Like the table groupings, the carpet allows the students to work on mathematics on their own, in pairs, as a small group, but also, as one large class. This space, the round table, and the cluster of desks provide useful places for the students to work with mathematical materials.

Physical mathematical work at the chalkboard

On the other side of the desk area is the smaller of two carpeted spaces. Jennifer created this one so that she and her students could use the chalkboard, overhead projector, and pocket chart during mathematics lessons. All of the equipment is positioned so that they are at appropriate height for the children to access. This place of gathering is not so much for active, physical investigations of mathematics, but for the class to come and record, display, or bring key mathematical ideas into focus for further studies.

In the big room

On the playground

Finally, Jennifer uses still other physical spaces to extend the children's mathematical learning beyond the walls of the classroom. They include an adjoining classroom that provides a large, completely open area in which to work, the children's homes, the natural environment, the playground, and another elementary school. The latter four offer contexts in which the students can develop and explore mathematical relationships within and across different settings.

Materials Matter

Along one of the walls in Jennifer's classroom are three large shelving units: one with blue drawers, one with yellow, and one with red. Each shelving unit contains sixteen drawers that hold a variety of materials used for mathematics. The blue drawers contain mathematical games and puzzles.

The yellow drawers contain calculators and different kinds of materials for mathematical investigations. The class uses these as physical objects with which to think about particular mathematics or with which to express their mathematical thinking. They include such items as: as number cubes, tangrams, calculators, base ten blocks, double-sided counters, geoboards, multi-link cubes, pattern blocks, attribute blocks, and zaks.[2] There are also many found items that the children have collected from their homes and neighbourhood such as different kinds of rice, pasta, beans, seeds, stones, chestnuts, and buttons. Here, the students make use of these materials for purposes of sorting and classifying, estimating and measuring (i.e., lines, area, volume, mass), exploring other number concepts, and for doing computations.

There is a third set of red drawers that contain different kinds of paper: plain, lined, graph, dot, and construction. The class uses these for constructing or recording their mathematics.

Together, the materials and organization of the spaces for learning create a portrait that tells of an open, accessible place that nurtures the growth of children's mathematical thinking. The classroom as a physical and expressive form of the relationships that are rooted here, communicates the importance for mathematical interactions to be flexible and diverse.

An environment is a living, changing system. More than the physical space, it includes the way time is structured and the roles we are expected to play. It conditions how we feel, think, and behave.

~ Jim Greenman[1]

Conceptualizations of Time

ABSOLUTE, true and mathematical TIME,
of itself and from its own nature,
flows equably
without regard to anything external,
remains similar and immovable.[2]

Sure in its certainty
TIME lives its independent existence

SEPARATE and d i s c o n n e c t e d

created and affected
by
no person
no space
•
•
•
•
•
•
•
•
no-thing.

TIME does not exist with us or because of us
TIME is just — **out there.**

And even though

TIME is undetectable through human perception[3]

it is **REAL**.

And even though

TIME is empty... spaceless... invisible... and

s
h
a
pe
less

it is **REAL**.

```
TIME is exactly infinite...
it can't be taken a-p-a-r-t
or pressed together..........
it can't be disposed of ......
it just IS................
```

Trust it is there.

For time is just this: the number of a motion with respect to the prior and the posterior[4] —Time as relative — as linear —

straight as an arrow — fusing with events — figuring the

spaces — delineating relationships — defining moments —

stringing beads — one after another after another after

another — each as a middle that unfolds from a beginning

and towards an end — forward moving — this kind of time

— these kinds of spaces — can be measured — numbered —

separated — sorted — sequenced — fixed — located —

walking the temporal tightrope — PAST behind us — that

was then — PRESENT — this is now — FUTURE ahead —

— how much farther...?

keep going ....

Time... unfolding in its wholeness... recursive and cyclical ... time becomes with space...and as one ... they take place as one ... one ... times unfold in moments.[6] ... back and along the tightrope of singular pasts, presents, and futures ... inseparably, fluidly, indistinguishably ... we live in the changing present in which past and future take place now.[7] ... there are no beginnings, middles, or ends ... living takes place in the now and we think, act, and become, moment to moment.[8]

The Wind... whirls. Birds make their nests in circles... The sun comes forth and goes down again in a circle. The moon does the same, and both are round... Even the seasons form a great circle in their changing... Life... is a circle from childhood to childhood and so it is in everything.[5]...

Time never becomes long because it originally has no length.[9] blending and blurring any separation... time-and-space-and-us ...[10] come to form together... and ... in wholeness and simultaneously

Enacting Re-conceptualizations of Time

Creating a classroom that engenders a systemic sense of place for mathematics requires consideration of how time is understood, how it is enacted through one's teaching, and the impact it has on shaping the spaces where mathematics occurs.

The languaging effect that my unconscious linear image of time had on the structuring of my mathematics program and the kinds of learning spaces that were then actualized is a clear example of this. Distinguished by grade levels, set within clearly marked boundaries of a Monday through Friday, 11:20 am to 12:00 pm and September through June frameworks, I conserved a delineated sense of place for mathematics in the classroom as a means to an end. The mathematical content of each lesson picked up from where the last one left off and moved the students forward in a concrete to abstract fashion to the next preplanned lesson. Together, these lessons formed instructional units that in turn became the year's mathematics program. In the larger scheme of things, each of these programs served as component parts within the K-12 mathematics curriculum.

My linear differencing of time languaged me to view my responsibility as teacher as a matter of providing mathematics programs that functioned as links in a curricular chain and progressed the students in their mathematics from one year to the next. In doing so, a straightforward linear time-space of mathematics teaching and learning was created and conserved. Consequently, what my conceptualization of time did not language was the possibility for time-spaces to embody fluidity, nonlinearity, and recursion — critical qualities inherent in my new metaphors.[1]

What if time was enacted as nonlinear, flowing, and recursive?
How might it be occasioned in the mathematics classroom?
What role could it play in re-placing mathematics teaching and learning?
What kinds of mathematics might then emerge?

Raising Questions and Questioning the Answers
~~Linear~~ Time-spaces for Mathematics in the Classroom

Engaging in wondering WHAT IF...? *revealed all kinds of unnoticed linear and static time-spaces with which I furnished my mathematics program. Becoming aware of these and bringing them into question has not been unlike my experience in confronting and re-imagining new metaphors.*

ASSESSING THIS FURNITURE: ***Why do I organize my mathematics teaching into two separate programs?***

Honestly, I could not come up with any meaningful reasons for why I separated and taught two one-year mathematics programs. Certainly, I could say that organizing time in this way clearly defined what students were to learn in second and third grade mathematics. I could also say that it served as the base from which I could plan the program overview, units of instruction, and individual lesson plans. However, given the metaphors with which I was now working, none of these explanations of cut and dry efficiency made sense to me anymore. Categorically complicated[2] *and fixed, they were devoid of anything that assumes complexity or strives to be life-giving. Up until now, the issue of time merging with space had never been an important consideration of my teaching. It was not a part of my map of what it means to teach mathematics. Linear ways of structuring time and spaces in which to teach mathematics was what was taken for granted and ritualized by everybody — administrators, teachers, parents, and the students themselves. But now the issue of time was emerging as a difference of critical importance that would not go away and demanded my attention. Now visible through confronting my temporal ignorance, contemplating its incoherence in my teaching and compounding this with my desire to embed an ecological sense of place in the classroom for my students' mathematics, rethinking my enactment of time proved to occasion new patterns of difference for the coming school year.*

I (it's me, Jennifer) continued to teach second and third grade students but began preparing for the upcoming year of teaching by considering the kinds of temporal patterns necessary for opening teaching and learning spaces that were organic and generative. No longer wanting to teach mathematics in a grade-to-grade manner but in a way that focused on nurturing an ecological sense of place and the dynamic growth of the children's mathematical understanding, I made the decision to teach a two-year program as opposed to two separate one-year programs or a single grade class. By doing so, I was able to expand the time-space from ten to twenty months. In contrast to my image of teaching and learning as a linear time-space, I envisioned the first year of the program as enveloping and co-emerging with the children's previous individual and collective mathematical activity.[3]

The second year was too conceived as a living curricular system and one that would create further growth of the children's mathematics. The Pirie-Kieren model of growth of mathematical understanding was a critical part of the structuring of the program as it locates children's embodied mathematical knowings as being where all new knowledge develops and thus, the place where the two-year program would begin.

If time is to be enacted as nonlinear and recursive, then the space in which it is a part must also emerge as such. Thinking systemically about the growth of mathematical understanding necessitates teaching to be a dynamic and responsive activity. Teaching and learning as living systems come into being moment-to-moment, co-existing and co-evolving in relation to each other. Although the two are viewed as activities that cannot be prescripted, this does not assume that responsive teaching does not require anticipative preparation on the teachers' part or that it is a random activity. As praxis, teaching responsively means being attentive and mindful towards how one's teaching impacts and is impacted by the class' mathematical work.

> Envisioned as a map, a teacher may locate mathematical topics, concepts and skills as important landmarks for the class' learning... what cannot be sketched out in advanced are the actual paths on which the students will travel to get to the mathematical locations or the understandings they will establish when and after they come to these sites... it is impossible to predict the conditions of the ever-changing landscape... like paths, what the students come to understand or learn is rarely if ever, straightforward but instead, spreads out in several directions and

I identified important learning aims that included the goals of the Ministry[4] and other documents (e.g., NCTM, BCAMT)[5] for the second and third grades but instead of sequencing them or categorizing them according to grade levels, I marked these aims on my two year map as important mathematical sites to be explored. As I did this, I also considered contexts that would be open to many different kinds of mathematical investigations[6] and ones that the class might revisit later on in the year or even again in the second year. While I mapped out a

entails twists, turns and switch-backs... curricula, understanding, and learning
a curriculum, I was ever mindful that it was just that – a map; distinguishing
imagined in this manner... distinguish these as co-emergent phenomena that are
pedagogical and mathematical features for teaching, and that the actual
brought into being through the students, their teacher, and mathematical settings.
forms my teaching and the students' learning would take were yet unknown.

Because it was my intention for the program not to have a prescribed teaching sequence, it was important for the learning spaces to be ones that not only had an open flow in terms of mathematical content but also, ones that would encourage the reintegration and renewal of the children's mathematics. To do this, I created provocative themes by myself, with the children, and in collaboration with other teachers. These themes arose from the children's ongoing work and usually came in the form of curious questions or specific topics such as, *Who Are These Things We Call, Numbers?*, *Snowflakes*, *Number Gymnastics*, and *Mathematics About Me*. So instead of instructional units organized by particular mathematics such as addition, subtraction, or geometry, these themes focused my teaching and the students' work on exploring mathematics as a diverse and interconnected whole.

It was here that I realized how the integrity of this program would be compromised if I continued the ritual of scheduling mathematics into 40 minute daily intervals. Affirming my disbelief that children only have short attention spans, I was inspired by the stories of the Reggio Emilia schools and looked for ways in which I might expand and enable flexible time-spaces to exist in the classroom. I seeded the program with the idea of establishing a place that focused on students' mathematical growth. It made sense then, for these themes to occur in time-spaces that not only allowed for the children to work on ongoing projects but also ones which enabled the mathematics to shape-shift into different forms of mathematical languaging, to branch off, intersect, and flow from real life contexts into purely mathematical ones and vise versa. Instead of cutting up and inserting mathematics lessons into 40-minute times slots, I opened up larger spaces of time within the day and even entire school days.

For the past two months, the students and I have been working on a variety of projects and investigations that focus on their meaning making of numbers within different contexts. This week, the class watched the film Notes on a Triangle. Reviewing the film on several occasions and from different mathematical perspectives piqued the students' curiosities. The vantage points they marked out — all of which happened to revolve around the idea of 'three', established the specific mathematical spaces for the students' explorations. Inside these learning spaces, the students worked to develop understandings for the three-ness of a triangle, the three-ness of particular numbers, why three is considered odd and when it becomes even, the rhythmic pattern of three-four time, and the aesthetic value of three in creating visual art. Here, concepts of addition, multiplication, number theory, patterns, and geometry were brought into being through the class' engagement with the film.

These time-spaces contrast with previous lessons in which I clearly marked beginnings, middles, and ends (i.e., introduction, development, and conclusion). They serve as examples of how mathematics curricula can arise in forms that are generative and embody a sense of flow; where time and space are taken to be inseparable and give rise to one another within the contextual boundaries of the class' mathematical experiences. Also highlighted is how mathematical concepts emerge from the class in a way that assumes mathematics to be alive. In these lessons, the mathematics taken up were not explorations in the practice of dissecting and dismembering but quite the opposite. Studying the film as an entity — as a mathematical form in itself allowed it to be viewed as a source from which further studies could be investigated and connected to one another, keeping mathematics intact and a dynamic whole.

4 Rs of Recursion
Reflecting • Re-viewing • Relating • Renewing

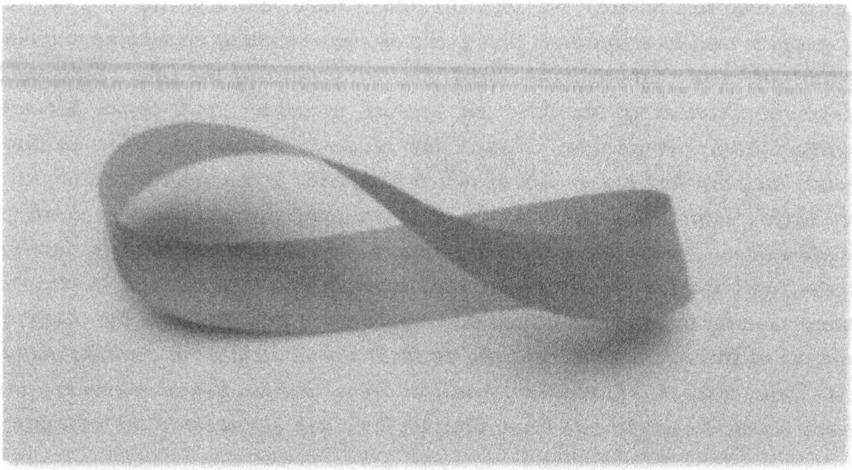

M. C. Bateson's description of learning as traveling along a Möbius strip[7] continues to be a powerful image that prompts me to examine and assess the relationships embedded in my teaching. It also challenges me to consider how I might consciously enable recursive forms of learning in the classroom. If mathematical growth is viewed from an enactive perspective,[8] like other forms of knowing, it is an embodied phenomenon that develops through relating and re-experiencing mathematics from an opposite side — a contrasting point of view, or seen suddenly through the eyes of an outsider. Because of this, students' learning is not achieved through repetitious acts of reproduction or sequential assembly lines of tasks. Doing so implies learning to be a matter of practising by redoing what one already knows or taking what one knows and adding to it in a piecemeal manner.

Mathematical growth as a recursive event connotes the actual changing of one's mathematical understanding in ways that are complex and co-emergent. While its evolution possesses qualities of self-similarity based on primitive knowings, what it becomes and what it occasions upon each recursion is something qualitatively rather than quantitatively different.

In this sense, recursion is not just a better understanding; it is a new understanding that is more than the sum of its parts. Mathematics teaching that provokes recursive forms of learning is more than C=B+A.

Instead of understanding as growing through an additive action, it is seeing the mathematics in a different way. Similar to the change in potential possibilities that occur as a result of structural coupling or drift,[9] *this is represented by the don't need boundaries in the Pirie-Kieren model*[10] *where once beyond a don't need boundary, one does not need the specific inner understanding that gave rise to the outer knowing. For example, a child who takes nine counters, arranges them into pairs, finds out that one is leftover, learns that this distinguishes the number as being odd and then repeats these actions to find out that eleven is also an odd number is a qualitatively different kind of understanding than that of a child who works with various numbers and then explains that "any number is odd if you divide it in two equal groups and there is one [whole] leftover." Through making meaningful images for specific numbers, the first child added to or thickened her understanding by performing the strategy of pairing up the counters and looking for one that was leftover to determine that eleven was also an odd number. The second child however, explored different numbers and identified a pattern or relationship that can be applied to any number to determine whether (or not) it is odd. This kind of sense-making of the mathematics is not repetitive or based on specific instances but conceptualizing the idea of odd numbers in a different and general way — for any instance. This child's thinking comes from a place of formalizing that is located within the Pirie-Kieren model beyond the don't need boundary which separates it from primitive knowing, image making, image having, and property noticing.*

So, what does this conception of learning mean for teaching in the classroom?

It places importance on making space for students to reflect on their mathematical patterns of thinking and to revisit their mathematics inside different contexts so that they may critique what they understand from their current place of knowing. Opening learning spaces like this not only allow for students to relate to the mathematics in multiple ways, but doing so also creates the possibility for occasions in which students reintegrate and thus re-new their mathematical understandings by conceptualizing them differently.

In wanting my students' learning to be recursive, one of my aims for the two-year program is to establish opportunities for these 4Rs[11] *to take place; to engage the students to reflect on, re-view, relate, and renew their mathematical understandings. This means that learning is not just an event that happens when one encounters something for the first time. It occurs when one comes at that something from an opposite side. Taking this image of traveling along a Möbius strip and situating it in the context of classroom mathematics where that something is mathematics, I am curious as to what an opposite side (or, sides!) might look like in the mathematics classroom, what forms it might take, and what kinds of understandings will arise.*

Making Three Spaces for Recursion

Can You Guess Our Mystery Number?! [1] was a mathematical gift of sorts, created and exchanged with the class' math peers at a different elementary school. This project brought together several weeks of class investigations in which Jennifer focused on the students' development of different images[2] for thinking with and expressing numbers through the language of: manipulative models, pictorial, informal or formal symbols (verbal or written), and real world situations through dramatization and descriptions. The children worked with a partner to craft a set of clues that would become the riddle for their chosen mystery number. Accompanying this list of clues, the students had to also provide eight other numbers in addition to the mystery one as possible choices from which their math buddies could identify the mystery number.

Playful as this project was, Jennifer was careful in designing it so that the structure of the activity encouraged the students to reflect and relate their conceptual understanding of number. By engaging the children in both the making and receiving of the mathematical gifts, Jennifer opened three potential spaces for her students' recursion – first, to reflect on their understanding of number, second, to create a riddle by relating their thinking, and third, to identify their math buddies' mystery number by using their understanding in a problem solving manner.

Before the children set off on the riddle-making project, Jennifer gathered the class onto the carpet and asked them to think about the number investigations they had done and what they now understood about numbers. She explained to her students that by collecting these images and recording them as a web on the chart paper, they could then use these to craft their clues. As the children volunteered their ideas about specific numbers, Jennifer wrote them down on the chart paper. For those who found this challenging, Jennifer encouraged them to think about a specific number and then fold back[3] and pick out a particular image that would be true or characterized that number.

After the class shared a few examples and non-examples for certain numbers, Jennifer prompted the children to see if they could take their specific number image and property notice[4] or express it in a more formal manner; as a descriptor for any number. Here, the children produced a variety of images and expressed that any number could be characterized in terms of: the number by which you could skip count up to it from zero or skip count down from it to reach zero, whether it fell between two given values on the number line, whether the number was divisible by a certain value, if it could be arranged into equal groups, if the number was more than (or, how much more than) or less than (or, how much less than) a certain value, whether it was considered to be even or odd, what the sum of the number's digits were, and characteristics of the number in terms of the place value of its digits. Once these had been recorded on the chart, Jennifer helped the students to find their partners and a place in the classroom so that they could get started on making the mathematical gift for their buddies.

Entering the second learning space, Mark and Danica left the carpet and went and found a table nearby to begin working on the project. After a brief discussion about the possible numbers that it could be, the two decided that 80 would be their mystery number. Taking a suggestion from one of their classmates, Danica and Mark thought it a good idea to first brainstorm and record as many different kinds of clues as they could think of and then choose the "*best*" ones for their riddle.

That afternoon, Jennifer asked the class to share the clues they had been working on and to discuss some of considerations they became aware of while working with their partner. She reminded the students to use the web of ideas that they had generated as a reference to help them critique their peers' clues regarding the variety and kinds of images being used. As well as giving clues that began with "The number is...", some partners had also incorporated clues that started with "The number is not..." or "The number does not...". When Jennifer commented that this to be a clever thing to do, the class agreed and they began discussing why using "not" in their clues encouraged a different kind of thinking for the person solving the riddle. Here, the idea of having to first understand the meaning of the clue and then having to be able to figure out what the opposite of it would be delighted the children. On returning to their tables, partners giggled to each other as they thought the process but in reverse so that they could craft these clues into their riddles.

Now satisfied with their store of clues, Danica and Mark set to making decisions as to which ones would be used for their riddle. It was here they noticed that some of their clues made it obvious what the mystery number was while others were less revealing. Mark and Danica sorted their clues and organized in such a way that their riddle opened with general characteristics about the number and then progressed to more specific ones. The two students also thought it clever to intersperse these with a few that were redundant or unnecessary such as "... *is less than 139*" after already giving the clue "... *is between 2 and 121.*" During another whole group discussion, the entire class agreed that the 'sum of the digits in the number' clue was the most telling of them all and that it should appear as the final clue. Finally, Danica and Mark moved on to generating eight other possible numbers for their riddle. They decided that each of the eight numbers should fit at least two or more of the clues and they were careful to select only those that satisfied the criteria.

"This way," Mark explained, "Tommy and Ewan will have to think [as to] *why* these aren't the mystery number."

"They might think the mystery number is fifty [because it's between 2 and 121, is an even number, is less than 139, does not have a 4 in the ones place, can be put into groups of 5, and it's more than 10] but it's not because its digits don't make eight", added Danica.

By the close of the afternoon, Mark and Danica's mystery number gift was ready to give to their math buddies.

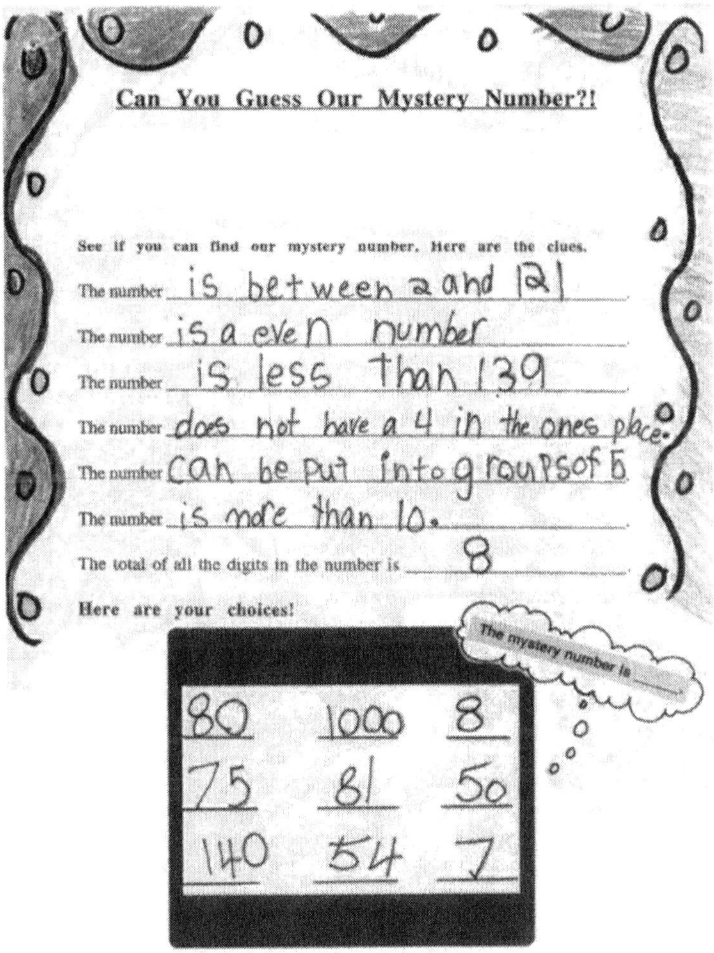

Can You Guess Our Mystery Number?!

See if you can find our mystery number. Here are the clues.

The number is between 2 and 121

The number is a even number

The number is less than 139

The number does not have a 4 in the ones place.

The number can be put into groups of 5

The number is more than 10.

The total of all the digits in the number is 8

Here are your choices!

80	1000	8
75	81	50
140	54	7

The mystery number is ______

Danica and Mark's mystery number riddle for Ewan and Tommy

One week later, Danica and Mark and, Ewan and Tommy received their mystery number riddles from one another. This third learning space that Jennifer created for this project required the children to take on a problem solving role. Not only was each set of partners to identify what they thought the mystery number to be, but they also needed to provide justifications for why the other eight numbers could not be the mystery number.

We think your mystery number is 80 because:

It couldn't be 1000 because it is more than 121.

It couldn't be 75 because it is an odd number.

It couldn't be 140 because it is more than 139.

It couldn't be 54 because it has a four in the ones place.

It couldn't be 81 because it can't be put into groups of five.

It couldn't be 7 because it is less than ten.

It couldn't be 8 because it is less then 10.

Finally, it couldn't be 50 because when you add 5 and 0 together, you get a total of 5 not 8!

Ewan and Tommy's response to Danica and Mark's riddle

In a different manner, recursion occasioned renewal of the students' understandings. These were times-spaces in which the students revisited a familiar context, reflected on the mathematics that had already been brought forth, and worked on connecting related yet different ways of thinking about the mathematics than they had done previously.

A Snowflake of a Different Kind

As part of the snowflake study that the class was investigating outdoors, in photographs, on video, and through stories, the children constructed their own Koch snowflake.[1] Each student identified an interesting pattern that they observed while creating the snowflake.

Christina's snowflake

As the students worked on their own to find a pattern in their snowflake, there was definitely an animate quality in the way the class perceived the mathematical features of the snowflake. Throughout the lesson, students were heard talking to themselves about how the patterns in the snowflake were "growing." An observation like this elucidates the conception that the students held toward their mathematics; it points to the fact that they did not see the purpose of the task to be one of locating the answer or as the retrieval of pre-existent static mathematics. Communicated in their verbal descriptions of growing patterns was the understanding that this activity was about keeping watch[2] for emergent mathematics.

Excited about the patterns they had found and fascinated that each one continued on "forever", the students decided to compose a short story that captured the mathematics they saw arising from the snowflake.

January: Julie's story is about a triangle turned fractal by its ever increasing number of sides.

The pattern of sides in Julie's snowflake

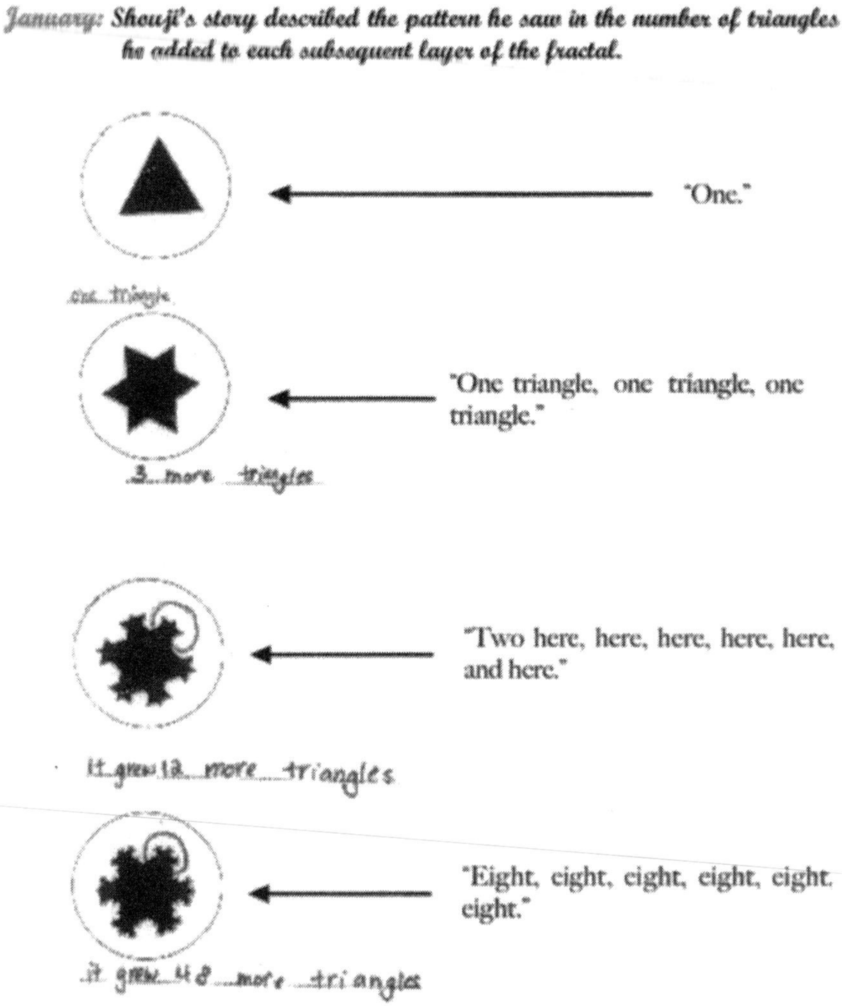

The pattern of triangles in Shouji's snowflake

January: ***Clare was fascinated by the "corners" that were emerging in the fractal. When asked to explain what she considered a "corner" to be, she defined it as an "elbow" in the snowflake; where the edge of the fractal "turned direction." Thus, the corners that Clare was counting were located on the perimeter of the fractal.***

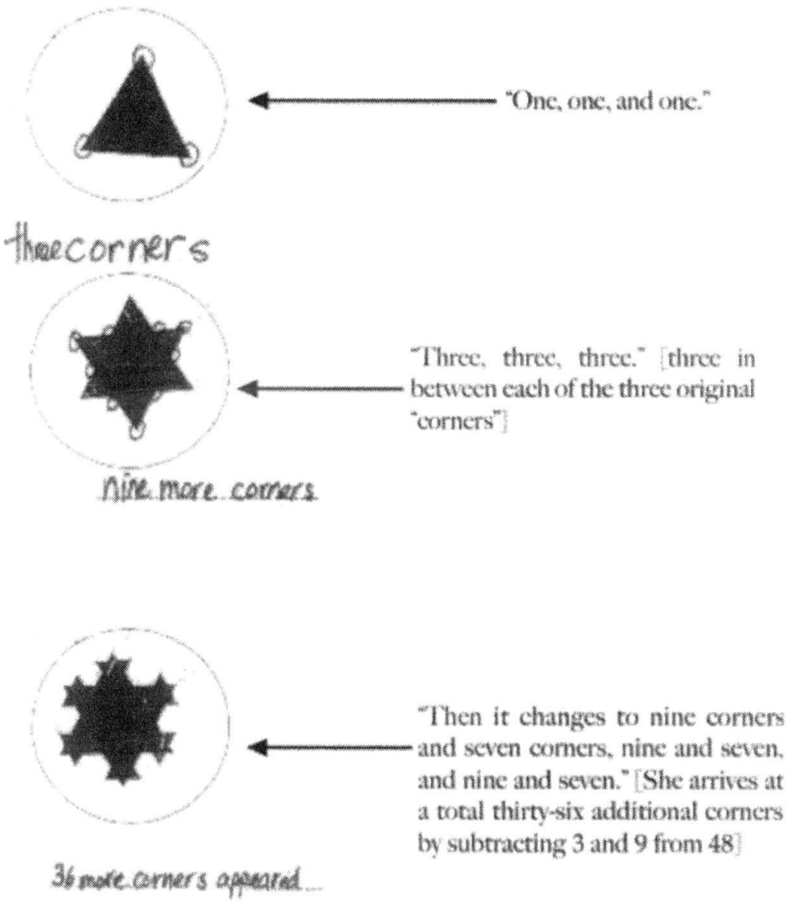

The pattern of corners in Clare's snowflake

The class revisited their snowflakes on three other occasions during the school year. For each of these sessions, Julie, Shouji, and Clare considered the total number of sides, triangles, or corners that they identified previously in each of their fractals. Re-viewing the first and then the subsequent layers of the Koch snowflake, the children worked to develop different ways of thinking and expressing the snowflake's growth incorporating their use of symbols and their knowledge of number operations.

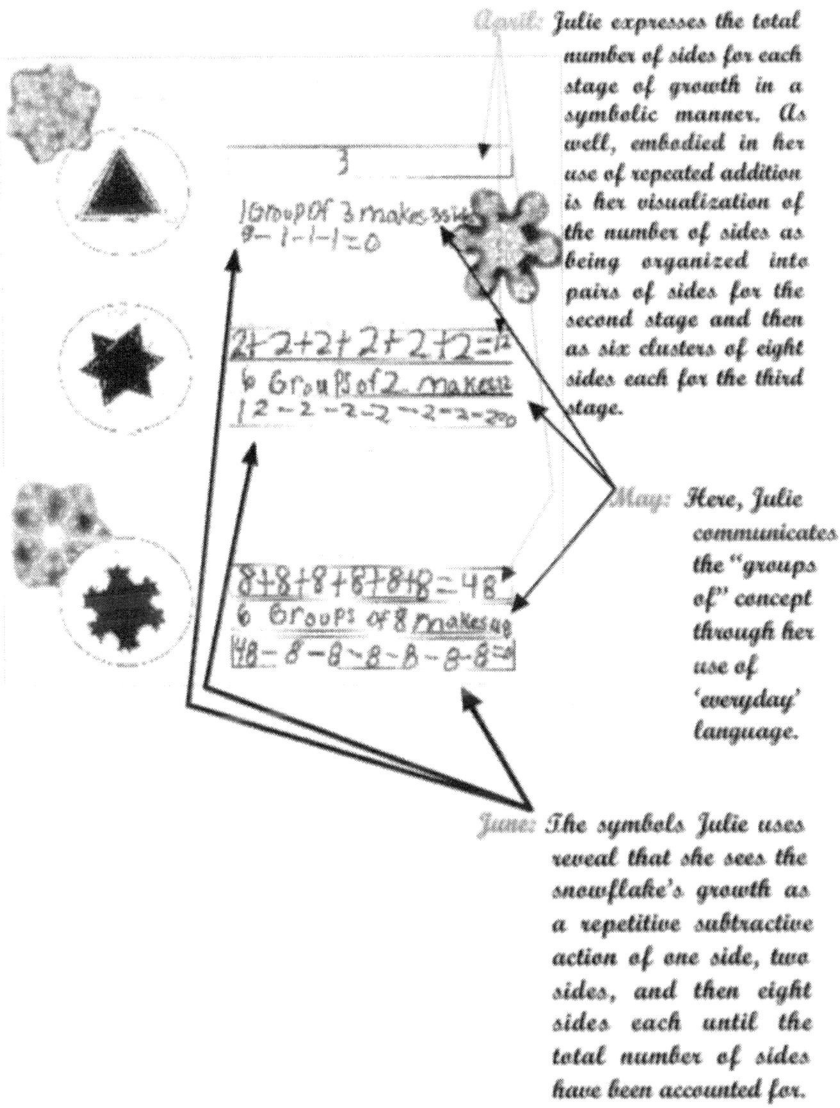

Three more interpretations by Julie

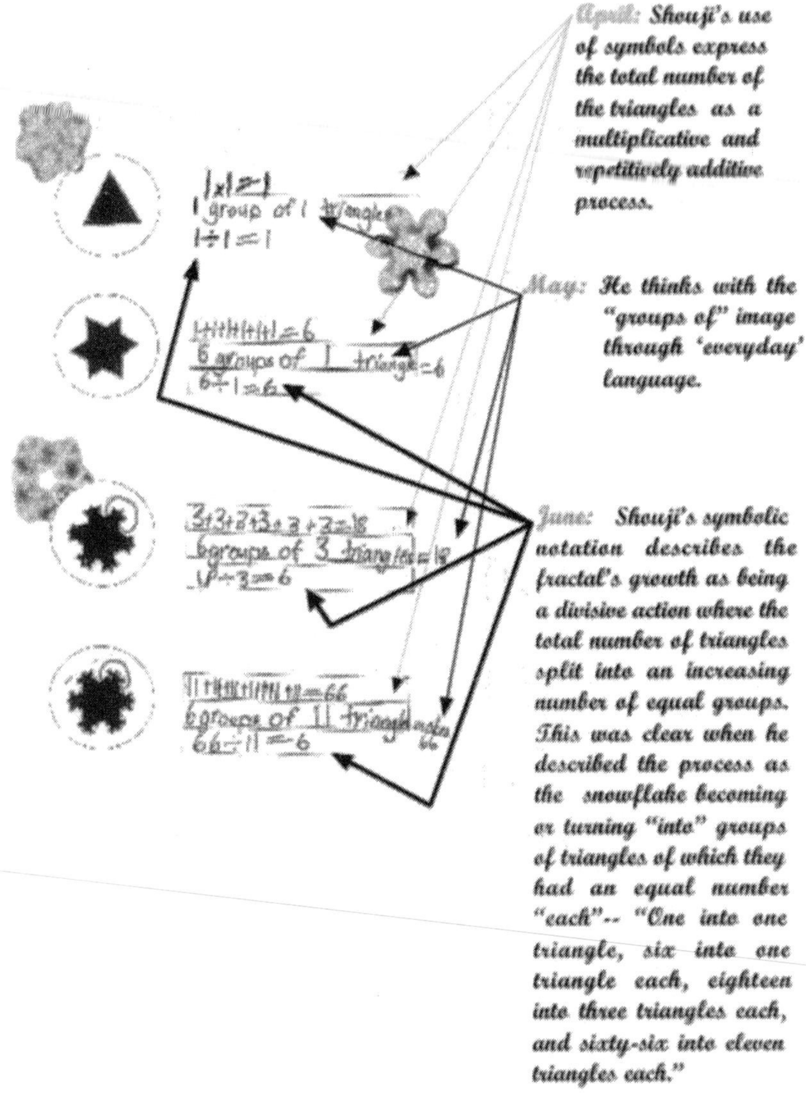

Three more interpretations by Shouji

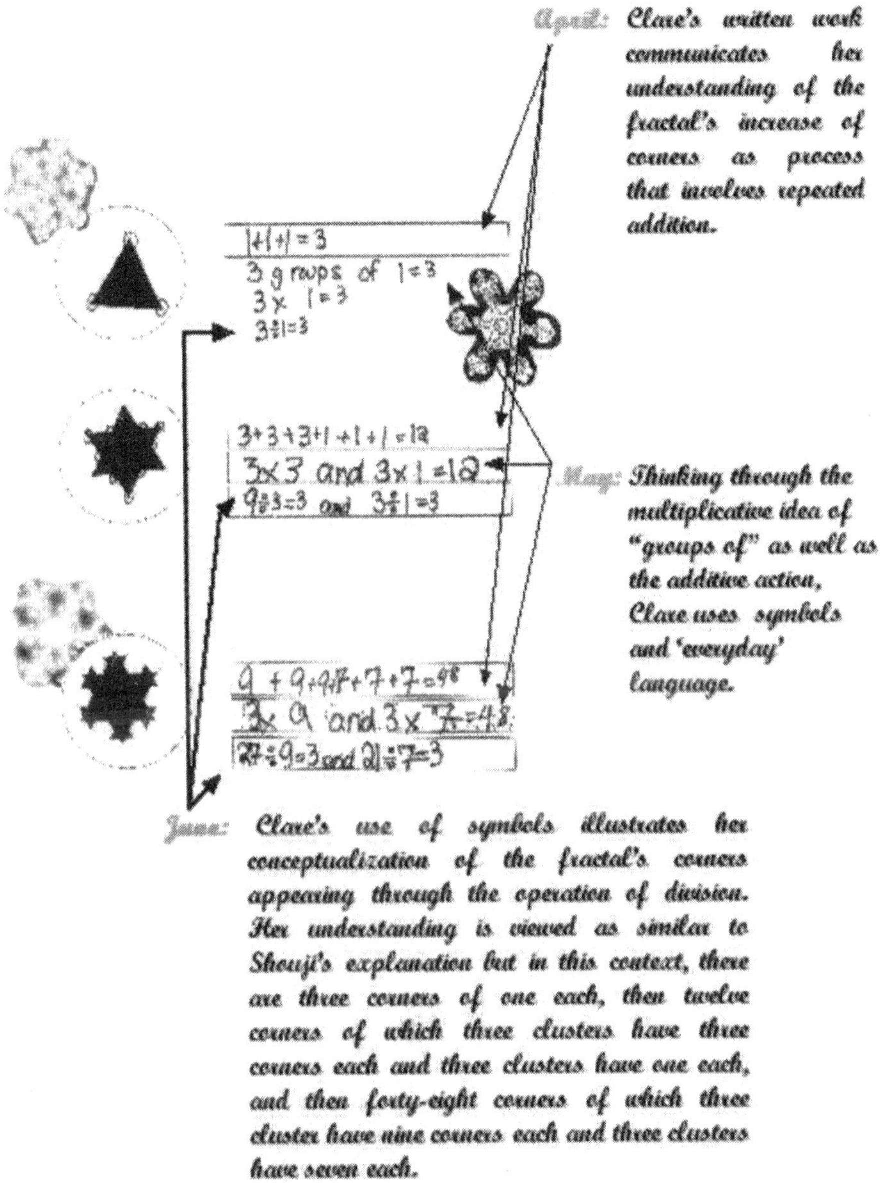

Three more interpretations by Clare

On still other occasions, the class returned to familiar contexts but investigated them in completely different mathematical ways than before.

Re-viewing and Seeing Differently

This Fall, the class watched the film, *Notes on a Triangle*. For the returning students now in the third grade, this was an opportunity for them to re view the film and for the new students in the second grade, it was their first time seeing it. Like the previous year, the students studied the film by watching it several times over but this time (and, in a very non-linear manner to that of last year) they saw new mathematics coming to life. As a result, the students laid down very different learning paths.

For example, some of the students noticed that the triangles in the film "weren't all the same." They watched the film again and this prompted the class to study the triangles more closely. The students used their fingers to make triangles, they drew them or cut them out of paper, and some went on a search around the room to gather them. Having a good collection of triangles, the class worked with a partner or in small groups to explore what was the same and what was different about them. Afterwards, the class shared their methods for doing this on the carpet. Some students explained that they studied the shapes by moving them about on their desks — sliding, flipping, rotating, and transposing one on top of the other to make direct comparisons about the lengths of the sides and the effect that this had on the "shape" of the triangle. Some students used rulers to measure the sides of each triangle while others took the shapes in their hands, turning and feeling each side, surface, and corner. Having gotten to know these triangles so intimately, the students naturally, gave them names.

"I call this triangle a[n] **almost all equel triangle** because only two of it's sides are the same."
~ Ethan

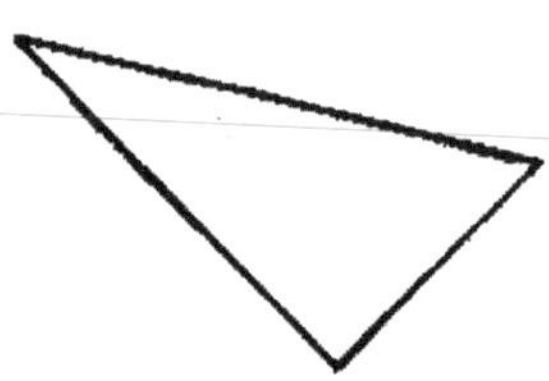

"I call this triangle the **deferent sided triangle** because it has 3 different lengths of sides."
~ Shane

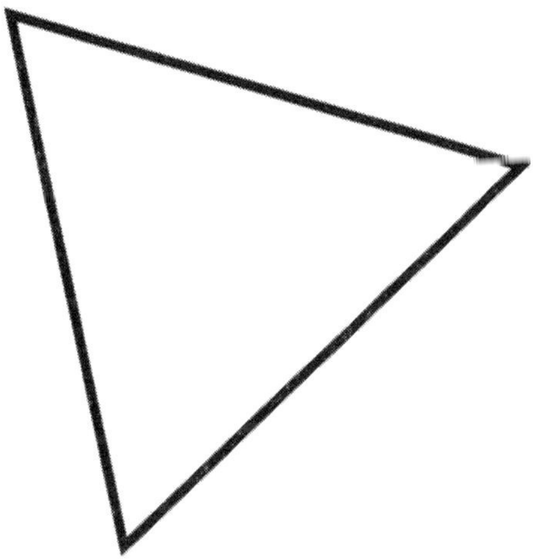

"I call this a **triple side triangle** because It has the same sides." ~ Steven

"They all have 3 corners [and] have 3 sidse." ~ Madelaine

"But they [differ in their] length of the sides." ~ Christina

What is the same about the triangles? What makes each of them different?

The class went on to explore how larger triangles could be created from smaller triangles...

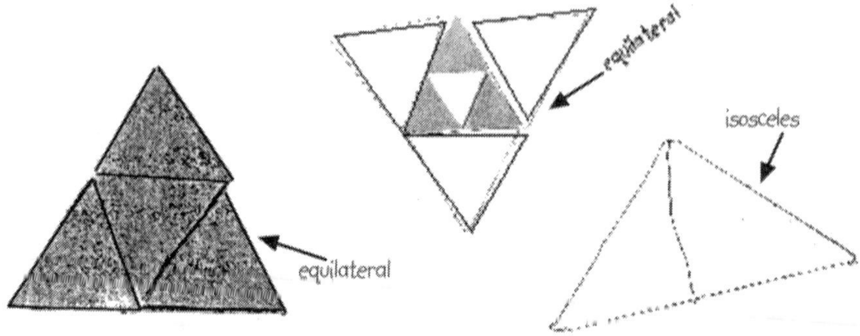

Annie's triangles

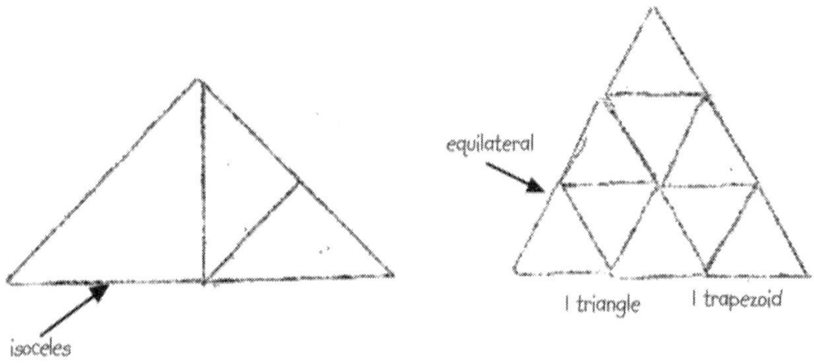

Christina's triangles

... and what other shapes could be made from triangles.

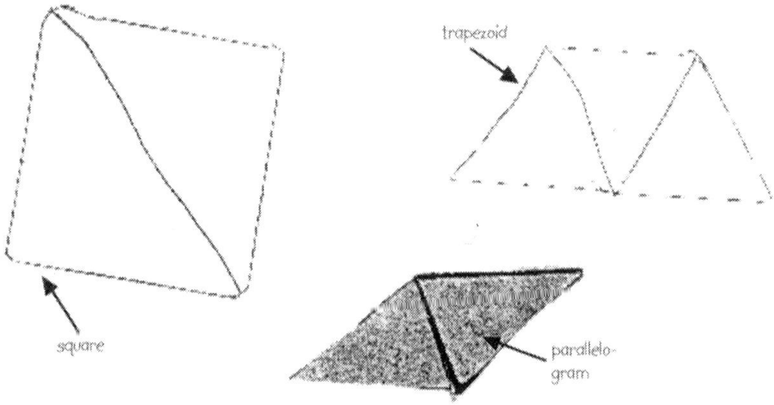

Robby's "other shapes" made from triangles

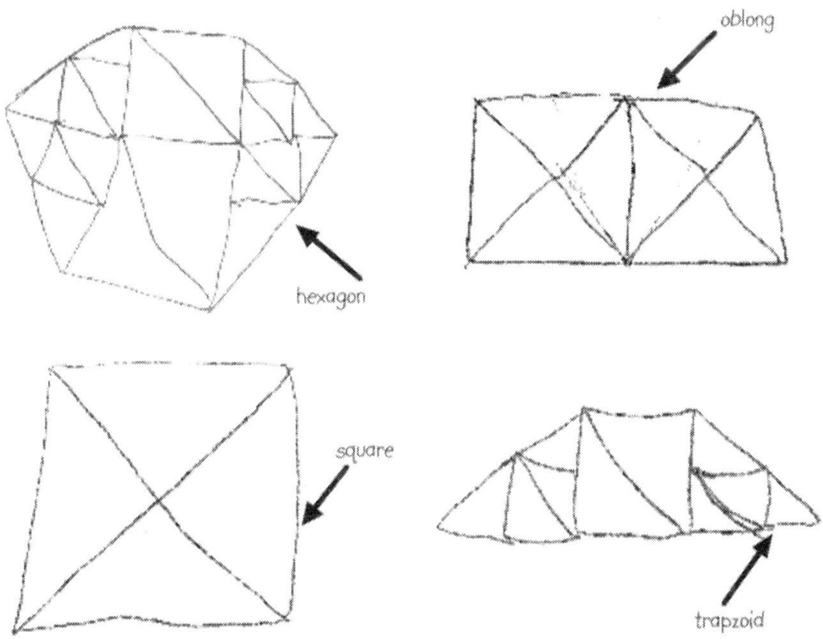

Mac's "other shapes" made from triangles

During the same time that the class was watching the film, five students had been working on another project that involved their construction of many different kinds of "pyramids" made from interlocking cubes. The children were sketching diagrams of the top, side, and bottom views of their structures when they told me that the bottoms of the pyramids looked like triangles:

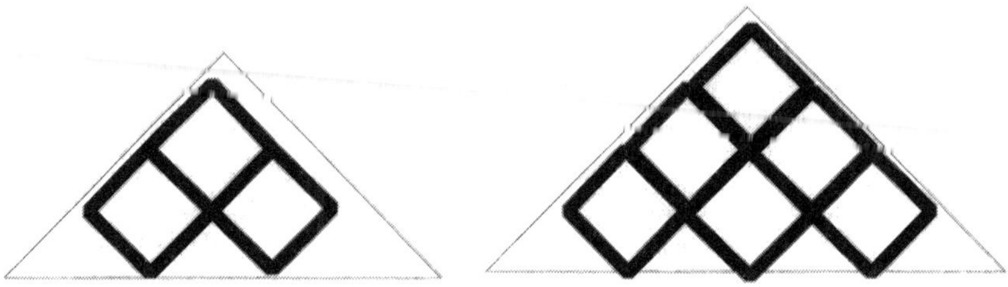

The group shared this with the class and this led into a study of how numbers too could be considered to be triangular.

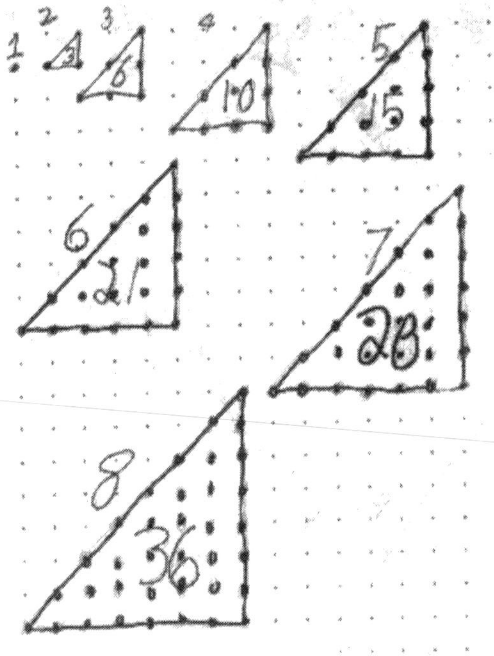

Timothy's diagram of triangular numbers and how much they increased each time

"I think the next number will be 45 because... 36 + 9 = 45 so 45 is the next one." ~ Timothy

"I think the next number will be 45 because 28 is 7 more than 21. 36 is eight more than 28. So 45 is 9 more than 36." ~ Steven

"I think the next number will be 45 because the pattern is odd, odd, even, even." ~ Clare

"The first has one dot. The second number has two [more] dots. The third you add three, the fourth you add four, the fifth you add five, the sixth you add six, the seventh you add seven, the eighth you add eight. The next one you add nine to it. So thirty-six... thirty-seven, thirty-eight, thirty-nine, forty, forty-one, forty-two, forty-three, forty-four, forty-five." ~ Clare

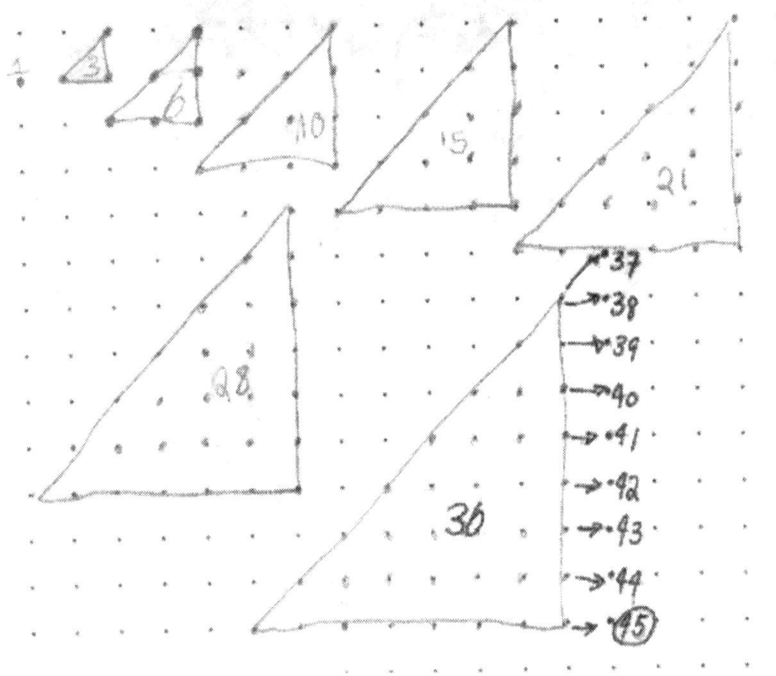

Clare's diagram which shows that the triangular numbers increase by a column of corresponding dots each time.

Recursion as Relations
When Triangles Become Square

(Three months after working with the film, *Notes on a Triangle...*)

> ***Last week, I introduced the class to the film, Dance Squared. Since then, we have been re-viewing it and using it as a source from which to occasion the class' further mathematics. The students worked together, posing prompts and exploring the geometries of the square. The students also identified square numbers — spurred on by their interest in triangular numbers. Reflecting on the class' curiosities and making plans for tomorrow's mathematics, I have decided to focus their work on both kinds of numbers. This will be a chance for the students to not only fold back and reflect on what they know about triangular numbers and square numbers but also, for them (on their own and with the class) to consider each in light of the other. Importantly, had the students' previous work with triangles*** **NOT** ***occasioned their investigation into triangular numbers, this lesson would neither have the same recursive potential nor be appropriate.***

(The next day)

Sitting on the carpet, in the middle of the circle, the children helped one another to build the first five triangular numbers with counters.

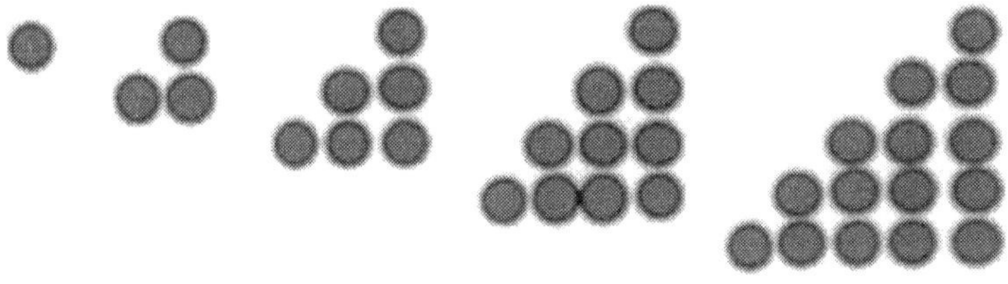

Triangular numbers built with counters

Underneath this, other students worked together to build a row containing the first five square numbers.

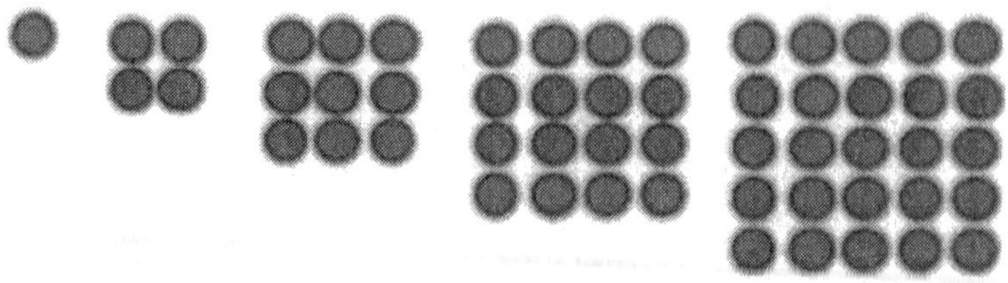

Square numbers built with counters

As they did this, Robby whispered something to Mark that in turn, started a very lively conversation between the two students. Robby and Mark then tilted their heads back and forth several times, all the while smiling and each exclaiming, "yeah!"

> ***I could not hear what they were talking about and so I asked Robby if he would share the conversation with the rest of the class. Robby flashed a bashful smile and then raised his voice to explain what he and Mark had been discussing.***

He pointed with his finger to the second and the third triangular number and then to the third square number and told me that, "that number and that number makes this square if you take it and flip it upside down and put it on top of it."

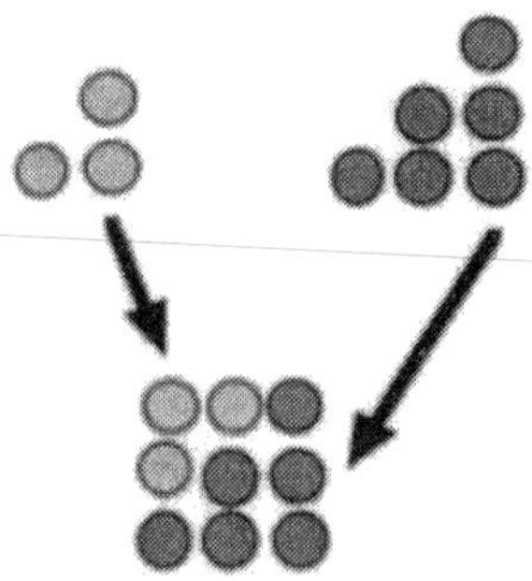

Robby's explanation of how the third square number is created from the second and third triangular number

> ***My "WOW!" and the boys' excitement for how they related the two number series together in a spatial way drew several other students into the conversation.***

Soon, the class began trying to make sense of this for themselves by talking with one another and displaying similar hand and body gestures to those of Mark and Robby.

> ***Wanting to maintain the focus and momentum of this investigation, I repeated what Robby had said but this time, I also built the numbers with counters as I spoke so that everyone could see the transformation that was taking place.***

"If you take the second and third triangular numbers and put them together like this," I said, demonstrating with the counters just as Robby had indicated earlier, "it is the same as the third square number."

> ***This prompted the rest of the class to continue the pattern by combining one triangular number with the one that preceded it in order to produce a square number. Not only did the class do this visually and verbally as Robby and Mark had done but also, arithmetically by adding the two values together, and physically by manipulating the counters of triangular numbers.***
>
> ***After the class produced the second through fifth square numbers from triangular ones, Danny reflected aloud and this time, he related the class' mathematical actions together into a connecting pattern.***

"It will be this number," Danny said as he pointed to the third triangular number, "but in the square number." Pointing to the third square number, which was nine, he explained that that was the resulting spatial structure and number when the third triangular number is combined with the second triangular number. Hence, six plus three makes nine.

I encouraged Danny to continue. Pointing with his finger in a left to right direction, beginning with the first triangular number that was one, then moving to the second triangular number that was three, and then to the second square number that was four, he communicated the relationship between the two series of numbers.

"This first number and the second triangular number will make the second square number. And the second and the third triangular numbers make the third square number. The third one and the fourth one makes the fourth square number and so on and so on..."

> *As I listened and watched Danny thinking about these numbers, I could see the recursive nature of his understanding — emerging from the mathematics already brought forth another connection between these numbers arose. Mark and Robby's observation which eventuated from their property noticing[1] that the second and third triangular numbers could become the third square number served as a place from which to begin our examination. The class' further collective work to apply this notion provided several more examples in which this relationship existed. Danny's understanding revealed yet another quality about the triangular and square numbers: that of a more formal, generative and predictive relationship between them.*

Two and a half years later...

Exploring the where and when of these recursions has proven to be dynamic and at times, unpredictable in nature — even fractal like. Sometimes the students' returning to or switching back on their thinking takes place spontaneously by surprise, say, inside of a conversation while other recursions are self-similar iterations that occur within a school year, inside of the two years or carried on even further. I have also learned that it is the mindfulness-awareness of the teacher that enables these recursions in the mathematics classroom to vary in form and purpose.

This process of time becoming a conscious part of my teaching not only reshaped the organizational structure of my mathematics program itself, it also had a profound effect on the kinds of learning settings that are emerging and the living forms of mathematics that are brought into being by the class. It has opened spaces for my students to interact with mathematics and in doing so, experience recursion as an exploration of the unfamiliar within the familiar.

What becomes clear to me now is the impact that one's conceptualization and enactions of time have on place-making in the mathematics classroom. How time is imagined and enacted directly structures the way in which it will be experienced by both teacher and students as well as what kinds of mathematical events and relationships are possible and impossible in the flow of the classroom.

I am also learning to let my ecological metaphors think me as I think with and in them. Doing so provides me with direction and focus to enact recursive notions of time in teaching mathematics. The forms of teaching and learning that have emerged so far are a stark contrast with my previous mechanistic ones that created a sense of place where students were expected to assemble specific forms of mathematics before moving on to more complicated lines of mathematical re-production. Instead, these new spaces for learning embody patterns of a temporal difference where past, present, and future exists all-at-once. Time is inseparable from space because it is defined by those spaces in which mathematical experience occurs. Time and space are not distinct but exist as one co-determining entity of time-space.

... children... are taught at a tender age that the way to define something is by what it supposedly *is* in itself, not by its relation to other things.[1]

~ Gregory Bateson

To know something is to know what that something is in the way it speaks to us, in the way it relates to us and we to it.[2]

~ Max van Manen

... the pattern which connects... How are you related to this creature? What pattern connects you to it?[3]

~ Gregory Bateson

Might that include this creature we call, mathematics?

Creating Patterns Which Connect

Interactional Spaces for Mathematics in the Classroom

Insight, I believe, refers to that depth of understanding that comes by setting

In thinking ecologically about the mathematics class, it is no longer possible for me to imagine learners as autonomous individuals. It doesn't make sense within a systemic mind-space for each learner to exist as a separate entity, acting on everything else as if everything else was part of the external environment. So even though I continue to recognize each student as an individual, I conceive the students to be individuals and simultaneously members of larger collective and environmental realms that are nested within one another.

In thinking more specifically about how I may nurture my students' mathematical growth, this work continues to involve making space for them to explore their mathematics as individuals. But rather than just furthering their understanding through a process of adding on new mathematical experiences, attention is also cast upon looking deeply and examining the understandings embedded within their mathematics. In this way, making opportunities for learners to not only engage in individual mathematical work but also, for them to consider how they understand the mathematics by drawing on their knowings and developing relationships amongst them. This particular space, then, seeks to occasion the emergence of the students' mathematics as they reflect on what they know by letting their understandings speak to one another.

experiences... side by side, learning by letting them speak to one another.[1]

Meeting with Mac

A Study of Opposites and Relatedness

Mac and I are sitting at a large table. Starting with any number of cubes, Mac is to find at least two possible ways to take away equal groups of cubes until no cubes remain and record each of the stories using whatever kind(s) of symbolic notation he wishes to use.

As Mac begins, I'm interested in the understandings Mac brings to the task. He hasn't had any formal lessons involving the operation of division or the ÷ symbol and so, I'm curious as to how he'll express the mathematical action of repeatedly removing a particular number of cubes. I'm looking for occasions in which to alert Mac to examine his understandings and engage him in thinking about how he might use what he knows to develop other ways of thinking about the mathematics at hand.

Mac begins with a pile of 10 cubes. "I started with ten." He writes the numeral, 10 onto a page in his notebook.

I take away two cubes.

Mac records -2 beside the 10 so that it reads 10-2. "Because we took two away so it's ten minus two."

I take another two cubes away and repeat this three more times.

Each time, Mac records and talks as he works, "minus two, minus two, minus two, minus two! And the answer is... zero!"

Based on what Mac's saying and how he's recording the taking away of groups of two, he views this activity as one of repeated *subtraction. He's comfortable and able to independently express the mathematical actions he observes using both verbal mathematical terms and recorded symbols. As well, these forms of mathematical language seem to be an integral part of Mac's thinking because he doesn't watch the events from start to finish and then formulate his equation, he does so in tandem—as they happen. In this way, Mac narrates the mathematical story as it happens.*

Pushing the ten cubes into the middle of the table, Mac starts again but this time, decides to take away groups of five cubes. "Ten minus five… ," writes 10-5, removes the remaining five cubes, "minus five…", "equals zero." He finishes his number story by writing down -5=0.

Having completed two ways to remove ten by grouping, I ask Mac to choose a larger number for the next task. He selects twenty to be the number and proceeds to add ten cubes to the ten that are already in the middle of the table. I then ask Mac if he knows of an amount that could be taken away in equal groups from twenty so that zero cubes remain.

"Hmm." Mac looks at the twenty cubes on the table. He places his elbow on the table and rests his head in his hand. Pursing his lips, he thinks out loud. "Maybe you could do… hmm…" Mac takes his elbow off the table and rests his hand down beside his notebook. He looks at his fingers and then stares across the room. Curling and uncurling the fingers of his right hand, Mac looks at me, "Hmmm… fives?"

I assume Mac's skip counting by fives to twenty on his fingers. To be sure, I'll ask him to explain his response.

"And how do you know that fives would be a good one to choose?"

"Because you can count by fives to any number." He demonstrates this to me by taking his pencil and touching it down on the table in two spots horizontal to each other with a good size space between them.

It appears that Mac's thinking with an imaginary horizontal number line along which he skip counts by fives to twenty. I'm not sure what he means by "any" number and so, I'm going to encourage him to continue.

"Like you can count by fives to twenty. Five, ten, fifteen, twenty." Now touching the two points but this time with his index finger, Mac skip counts by fives, pointing in a left, right, left, and right manner while also moving his head from side to side in a left to right motion — like a metronome, marking the numbers as he counts and points:

5	10
15	20

Demonstration of Mac's counting action

Now, it's clear to me that Mac is not counting along a horizontal line but between two points that separate the numbers into those that end with 5 and those that end with 0. His counting action might have arisen from the hundred chart that we use often and hangs in the classroom.

The numbers 1-100 are organized on this chart in a 10x10 grid. Because of this, they fall into columns where each number increases by ten as you move to the next one directly below it and so, all of the numbers within a particular column share the same last digit.

It's also possible that Mac's counting method could be a result of the rhythmic pattern that is generated from skip counting aloud.

"I started with twenty, minus five, minus five, minus five, minus five, equals zero." Mac records 20-5-5-5-5=0 into his book...

... while I take away the four groups of five.

"So, that works."

I gather the twenty cubes into the middle of the table again.

Mac begins the next number story. Moving the cubes with his left hand and recording with his right hand, he takes away groups of two cubes each. "Twenty minus two, minus two, minus two, minus two, minus two, minus two, minus two, minus two, minus two, minus two, equals zero. So that works." In a similar way to his other number stories, Mac records 20-2-2-2-2-2-2-2-2-2-2=0.

Up until now, Mac has used groups of two and five to divide the cubes of ten and twenty. He's doing this by applying the "opposite" process of repeated subtraction. In other words, through repeated addition Mac's skip counting to arrive at the target number and then simply reverses the process by transforming it into a repeated subtraction equation. Since skip counting by multiples of two and five appears to be Mac's strategy of choice, I wonder what other thinking he might bring forth if I ask him to continue working with the number twenty.

"Is there another way?" I place the cubes in front of Mac.

"Hmm... you could do it by tens." Mac quickly writes 20 down into his notebook.

"We could do it by tens." I take away a group of ten cubes.

Mac writes -10 to the right of the numeral, 20.

I take away the remaining ten cubes.

"Minus ten." He records -10 and then =0. "Equals zero."

I put the cubes back into the middle of the table.

An attempt to prompt Mac to think of yet another way to evenly divide the twenty cubes.

"Oh..."

"Try by fours."

An impulsive suggestion!

"Okay." Mac begins talking slowly and taking away groups of four cubes. Each time, he looks to see how many cubes remain.

Mac appears unsure. This is probably because four is not a number by which he usually skip counts and therefore it isn't a number he associates with twenty.

Recording as he works, Mac completes the equation, 20-4-4-4-4-4=0.

Pointing to what Mac has recorded, I bring attention to two of his equations. "Now, looking at what you got here for twenties... look at the twenty give away groups of five [20-5-5-5-5=0] and look at the twenty give away groups of ten [20-10-10=0]. Do you notice anything about the number of groups, Mac?"

With his pencil, Max points to 20-10-10=0 and 20-5-5-5-5=0. "That this is just double of this".

"Why is that?"

Because there's two fives in ten so that's four." Mac then demonstrates this by pointing to each "5" in his equation. "There's FOUR fives there and TWO tens and one twenty. This is FOUR fives... TWO tens and one twenty."

By inviting Mac to stand back and reflect on the two number stories, he identifies the two equations as being related.

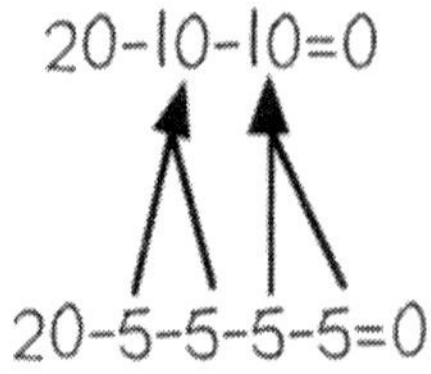

Mac's explanation of how 20-10-10=0 results from 'doubling up' four groups of five

Turning Mac's attention to the other two number stories (i.e., 20-4-4-4-4-4=0 and 20-2-2-2-2-2-2-2-2-2-2=0), I ask him what he can tell me that is the same or different about them.

"There's just double twos as there are fours... Pretend I didn't write that [points to 20-2-2-2-2-2-2-2-2-2-2=0] down and I just counted this [points to 20-4-4-4-4-4=0] — how many fours. There are, one, two, three, four, five. There's five fours. So then, I would just have to double that to see how many twos there'd be if I did twos. So there's ten twos."

Intriguing. As Mac explains how he identifies ten groups of two as being in five groups of four, I get a real sense of the flexibility of his understanding. Taking his previous idea of 'doubling' as a process of multiplying by two, Mac's thinking in this context effectively shape-shifts the process by inversing it. In essence, the doubling now becomes 'halving'.

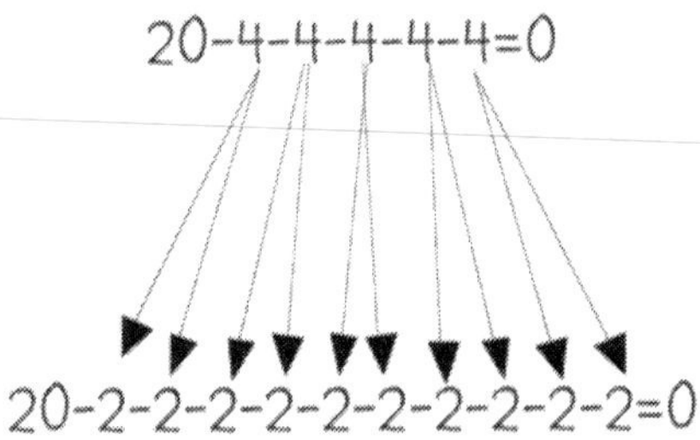

Mac's explanation why there are twice as many groups of twos than groups of fours

Smiling at Mac, we continue. "Okay, let's see if we can use some of that as we go along."

Mac nods his head. "Yes!"

"Pick another number."

"Twenty-four."

I add another four cubes to the pile that is already in the middle of the table.

Mac jumps in and quickly splits the pile of cubes in half or, into two groups of twelve. "Twenty-four minus twelve minus twelve." He records 24-12-12=0. Mac explains that he doesn't need to count the remaining pile because he "knows" that it will also be twelve.

Again, I want to prompt Mac to reflect on his thinking and consider how it might impact his understanding now.

"... taking that", I point to 24-12-12=0, "can you think of another number story that you could tell?"

Mac looks at the equation. "Groups of two."

"And how many twos would you have?"

Mac examines the equation again and then down to his hands that he is holding palms up. However, they cannot be seen because they are hidden underneath the tabletop.

I tell him to bring them out!

Mac skip counts by twos and keeps track by unfolding one finger at a time until he has six fingers extended. "Two, four, six, eight, ten, twelve. So double six [i.e., groups of two] is... twelve. So I'd have twelve twos. Okay." Mac sets to recording the number story, "Twenty-four minus two, minus two, minus two, minus two, minus two, minus two, minus two, minus two, minus two, minus two, minus two, minus two, equals zero... twelve twos." 24-2-2-2-2-2-2-2-2-2-2-2-2=0.

Mac's carrying forth his understanding from the last task and applying it here to help him determine how many groups of twos are in twenty-four. By making use of the equation that divides twenty-four into half, Mac takes one of the twelves and skip counts by twos to find out that there are six groups of twos in twelve. Then, he simply doubles the number of twos to get twelve groups of two. Like his other calculations, none of these are done with paper and pencil but through verbal and physical forms of computation.

"So, what's another way that we could make..." I put the twenty-four cubes back into the centre of the table.

"We could do... try fours!" Mac reaches for the cubes to begin taking away groups of four.

I'm not sure why Mac is reaching for the cubes. Is it because he thinks that since twenty-four is divisible by two that it'll also be divisible by four since four is double of two and he wants to check this with the cubes?

I ask him to estimate how many fours he thinks there will be.

"Hmm..."

Given that Mac doesn't guess or estimate, I have to take his idea of groups of fours to be a conjecture and not something that Mac knows as a fact. I want to help Mac to develop his thinking by connecting something he does know to solve for what he's trying to figure out. So instead of letting Mac continue to use the cubes in what I suspect will be a strategy of trial and error, I'm going to interrupt what he's doing and suggest that he reexamine what he's already done.

"How could you take something you know here..."

Mac moves from the cubes to his notes and uses his equation of 24-2-2-2-2-2-2-2-2-2-2-2-2=0 as his jumping off point. "I could say... I would just take half of these off," pointing to the groups of twos, "because I know there's twelve of these." Mac covers up six of the groups of twos with his hand so that they are no longer visible and explains to me, "So take half off twelve twos." Through Mac's actions, he demonstrates the need to take away six twos in order to form the groups of fours with the other six twos. "... so just count by twos." Mac proceeds to count the remaining twos. He begins to skip count them by twos and then realizes that he isn't wanting to find out how much remains but rather, how many groups of four there are in twenty-four. He switches his strategy. This time, he takes his pencil and partitions off two twos at a time while he counts the number of fours. "Okay, that'd be one, two, three, four, five, six. So there'd be six... six fours. So one, two, three, four, five, six." Mac keeps track of the number of '- 4s' he's writing to produce 24-4-4-4-4-4-4=0 on paper. Finally, Mac checks his work by taking the pile of 24 cubes and physically dividing it into six groups of four cubes each.

"Okay, taking what you know from the fours, can you do something new?"

"I could do... hmm..." Mac looks at the cubes in front of him and then looks at me. "Groups of eight?" Mac starts in and implements his doubling up strategy. "So just...", reaching for the piles of four cubes, Mac pushes two clusters together until they are all paired up and are now in groups of eight. "And there'd be three groups." He picks up his pencil and records 24-8-8-8=0. "... minus eight, minus eight, minus eight, equals zero. There we go!"

"Okay. Is there any other way we could share twenty-four?"

Mac shakes his head. "Hmm. I don't think so." He looks over the three groups of eight cubes in the middle of the table moving his eyes and head in a left to right direction. "N...nope!"

I push the cubes back together into a pile of twenty-four.

I'm attempting to get Mac to start anew and think with a different image[2] than groups of eight.

Mac looks at the cubes and sputters out, "one" softly. He takes single cubes away, "You could do one... minus one... so twenty-four minus..." Using his left hand to move the cubes away and his right hand to record as he acts and tells the number story out, Mac monitors the number of ones being recorded in his equation. He produces 24-1-1-1-1-1-1-1-1-1-1-1-1-1-1-1-1-1-1-1-1-1-1-1-1=0 "And the shortest way of sharing is one group of twenty four." Mac records 24-24=0 down into his book.

Once again, I'm going to ask Mac to review what he's done. Because he's been working exclusively with the images of splitting in half and doubling of the numbers, they don't allow for six or three to be possible values by which twenty-four can be divided. I'll try to open another space for him to work.

"So if you were to partner up the number stories that you were able to figure out from each other... So this one here" [pointing to the 24-1-1-1-1-1-1-1-1-1-1-1-1-1-1-1-1-1-1-1-1-1-1-1-1=0] and then to [pointing to 24-24=0] — the number story Mac had written as the reverse or the opposite story to the previous one]... giving away one whole group... connect them with arrows, what other ones did you..."

"Well, I first did a group of twelve and then from that twelve, I decided to do groups of twos, and from that two, I decided to put them together and I did by fours, and then I just added two fours together and I did eight! So those are all connected." Mac summarizes and connects the number stories with arrows in that order. He also connects 24-2-2-2-2-2-2-2-2-2-2-2-2=0 and 24-4-4-4-4-4-4=0 as related "because... four and two... because two twos make up four." Rather than keeping to the number stories about 24, Mac moves on to all of the other equations he has written. He relates 20-5-5-5-5=0 to 20-10-10=0 and 20-2-2-2-2-2-2-2-2-2-2=0 to 20-4-4-4-4-4=0. Framing the equations in terms of doubling and halving, he explains that he doesn't see 10-2-2-2-2-2=0 and 10-5-5=0 as being related and so he leaves them as they are.

Instead of trying to bring him back to twenty-four, I shift the conversation to exploring one last number.

Mac tells me to pick a number that can be divided into two equal groups.

Given what Mac's said and my intent to engage him to think still differently, I want to choose a number that'll divide evenly by two but result in an odd composite number.

Choosing eighteen, I ask him if he thinks it's an appropriate choice.

He pauses for a moment. "Yes". With the eighteen cubes in front of him, Mac sets to work. "Okay!" he exclaims. "Let's do…" Mac stops again. He places his hand to his lips as he looks at the cubes.

"What do you know about eighteen?"

Another pause. "… How about nines." Mac playfully slaps his hand down on the table.

"What's nine about eighteen?"

"There's two nines in eighteen."

Again, he's using the image of doubling or halving of a number.

He writes 18-9-9=0 onto the page. To this, he adds what he calls, the "shortest" number story. "We could also do eighteen minus eighteen equals zero" and writes 18-18=0. He follows this with the "longest" number story. "And we could also do eighteen minus one…". Mac talks out loud as he records this on paper and eventually produces 18-1-1-1-1-1-1-1-1-1-1-1-1-1-1-1-1-1-1=0.

I ask Mac to review the three number stories he's written for eighteen. "Now looking at any one of these, can you make up something new?"

"I could... this." Max chooses to rework 18-1-1-1-1-1-1-1-1-1-1-1-1-1-1-1-1-1-1=0.

"How do you know that?"

"Because there's two nines in one eighteen and we could... and this" pointing to 18-9-9=0, "is opposite."

Here, Mac describes how 9+9=18 and 18-9-9=0 are inversely or "opposite[ly]" related.

He points to 18-18=0 and draws an arrow so that now 18-9-9=0 is connected to both 18-18=0 and 18-1-1-1-1-1-1-1-1-1-1-1-1-1-1-1-1-1-1=0. "So you put those together too". Mac looks at me and concludes, "so... that makes it."

Mac appears to be finished this game because he's related all of the number stories to one another.

I challenge Mac to see if he can find another number story. Focusing his attention on the string of "ones" in the equation 18-1-1-1-1-1-1-1-1-1-1-1-1-1-1-1-1-1-1=0, I invite him to, "see if you can find some more groups in there."

"We could do groups of two."

Again, putting his doubling strategy to good use!

I acknowledge two as being a possibility for another number story and encourage Mac to make sense of it. "How do you know you can do groups of two? What tells you that's possible?"

"Because... you can... eighteen is an even number."

"And that means...?"

"As an even number, you can use it and put groups of two in."

"Okay."

Mac takes his pencil, "Okay, eighteen minus two, minus two, minus two..." and begins to record his number story. Mac continues writing, while keeping track by skip counting the twos he has written aloud "two, four, six... sixteen, eighteen." "Equals zero," he says as he finishes the equation, 18-2-2-2-2-2-2-2-2-2=0. Mac tells me that he can show that nine twos makes eighteen by "just counting the ones and see how many twos there is."

Mac's using the same strategy he developed to determine how many fours there are in twenty-four.

"We can count by two ones at once. One, two, three, four, five, six, seven, eight, nine." Mac counts the nine twos that are made by pairing up two ones at a time. "And one, two, three, four, five, six, seven, eight, nine." He counts and points with his pencil to 'prove' that there are indeed, nine twos to make eighteen in his equation. "Now, we can put these two together too." He draws an arrow to connect 18-2-2-2-2-2-2-2-2-2=0 with 18-1-1-1-1-1-1-1-1-1-1-1-1-1-1-1-1-1-1=0 to indicate the relationship of doubling the single units or ones of this even number in order to produce a number story that has groups of twos in it.

Pointing to his first number story, 18-9-9=0, I ask him to look at the "*nine and nine. Can nine be told another way? In terms of groups?*"

Mac looks down at the equation and thinks out loud. "You can tell it as... three groups of three! Or we can tell it as... groups of nine." Since he's already recorded eighteen shared by two groups of nine, he sets to work on determining how many groups of three there would be.

"*What would that look like for eighteen?*"

"For eighteen... that would be... there would be six threes in eighteen because there's two nines there." Mac points to 18-9-9=0 in his notebook. "So two nines in eighteen. So one of those two [nines]. There would just be... let's just pretend each of these are three [i.e., three threes] so that would make six." Mac points to the first nine and then the second nine in his equation. "So six groups of three... so six groups of three!" Mac goes to write this down in his book but stops and looks at me.

Here, the idea that Mac's thinking with — "six groups of three" isn't the same idea that he imagined his other equations with.

I ask Mac how he could reinterpret the story by telling it from the perspective he had expressed the others.

"Eighteen, minus three minus three, minus three... equals zero." Mac records 18-3-3-3-3-3-3=0 into his notebook.

An opportunity for Mac to bring together the image of repeated subtraction that he's been working with and the image he already had for multiplication.

To do this, I repeat something he's said previously. "Now Mac, did you hear what you said? You said, I HAVE SIX GROUPS OF THREE."

"Yes..." Mac continues to look at me, waiting for me to make my point.

"What operation do you think of when you hear GROUPS OF?"

"I think of... I think of... like say, I wanted groups of four." He takes the eighteen cubes and arranges them into four groups of four cubes and a group of two. "A group of four, a group of four, a group of four, and a group of four, and a group of two. Those are groups."

"Okay." I push the counters back into a pile and pick up Mac's pencil and hand it to him. "How would you write six groups of three?"

He takes the pencil, and says, "six" and then stops and asks me, "Should I use symbols or write it [i.e., in words]?"

I'm curious what symbols (e.g., + or x) or words (e.g., "and" or "groups of") Mac might use to express how he's thinking about six groups of three.

I tell Mac to record any way he wishes to express his thinking.

"Six times three equals eighteen". Mac writes 6x3=18. He takes a look at this on the paper and then explains to me that it does not make sense with the game we are playing or the other equations he has recorded; that the object of the game is to TAKE AWAY the number in equal groups until there is nothing left. "So that wouldn't work because we're trying to make it equal zero."

Now here's a chance for Mac to consider the complimentary relationship between the mathematical forms of 18-3-3-3-3-3-3=0 and 6x3=18. After all, it's this mathematical back-and-forthing that Mac's been doing with his actions of doubling and halving. He's been ultimately determining how many equal groups will make up the given number. Once he's done this, he then REVERSES his thinking so that it moves it in the opposite or complimentary direction. He does this by expressing the number as being taken away in equal groups until zero is reached.

Mac studies the two equations again. This time, he does not consider them as being unlike each other, but instead, as both communicating eighteen as six groups of three. The difference, he explains is that with 18-3-3-3-3-3-3=0 as opposed to 6x3=18, "we are not plussing" [the groups of three] "we are minussing them.. When you are taking away it would be subtraction. It's just like this [pointing to the multiplication symbol in 6x3=18], but it's subtracting not adding."

In his explanations, Mac not only identifies the inverse relationship between repeated subtraction and multiplication but also, how multiplication can be interpreted as (repeated) addition.

"They're opposite but they are also related." Mac draws an arrow to connect the two equations together.

"Now, what do you mean by they are opposite but they are also related?"

"Because there are six groups of three. Well, they are not really opposite because there's six times three and there's six groups of three here except for, this is plussing and this is minussing."

"So that [i.e., plussing and minussing] is opposite then?"

"Yeah but they are also related because there is six groups of three here except we're just taking them away. We're not plussing them." Mac reaches for the cubes on the table and shows me a group of three being added six times to make eighteen and then, in an "opposite-related" manner, how six groups of three can be taken away.

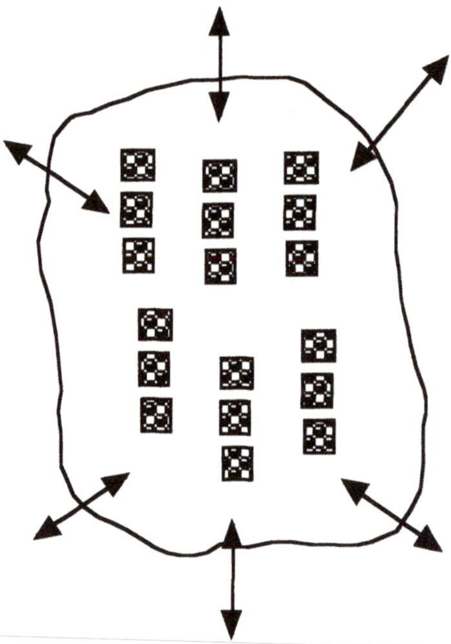

Mac's demonstration of 6x3=18 and 18-3-3-3-3-3-3=0

"Knowing what you know about this [i.e., the opposite relationship of 18-3-3-3-3-3-3=0 and 6x3=18], can you think of any other possibilities?"

Mac employs his doubling strategy one final time to solve for the last number story. "I could do three groups of six because there's two threes in a six. And I just counted by twos." He shows me that by pairing up the threes in 18-3-3-3-3-3-3=0 is the way he can check to make sure that the groups of three will double up without any remainders. Mac counts the number of groups of six that can be made from the six threes. "So one, two, three. So eighteen minus six, minus six, minus six, equals zero". He records 18-6-6-6=0 into his notebook.

"So you did this one [pointing to 18-3-3-3-3-3-3=0] and then you found out the oppositely-related one [pointing to 6x3=18] — six times three..."

"So, so this one would be related to this one too." Mac draws an arrow to connect them together.

"Do you think you could think of a multiplication sentence that would be the opposite of this one?" [pointing to 18-6-6-6=0]

"Hmm..." Mac pauses for a moment. He rewrites the equation but replaces the subtraction symbols with ones of addition. "Eighteen plus six plus six plus six equals zero."

I attempt to retell and act out the equation that Mac has written using words and my hands to show the additive action. "So if you had eighteen things and then you add six and add six and add six, you would end up with zero."

"No. That doesn't make sense."

I suggest that we use the blocks to try and make sense of it.

Mac agrees.

I mirror the same actions that Mac demonstrated when he showed me how adding and taking away six groups of three were oppositely-related. "So we did eighteen minus six, so I'm doing the action, right, you talk it out."

Mac chimes in as I take away the groups of six "... minus six, minus six... zero."

"Now, let's do the opposite." Still keeping the eighteen blocks in three groups of six, I perform the opposite action of bringing back the three groups of six so that they form the original pile of eighteen. "So if we go..."

"Eighteen plus six..."

I stop what I am doing and return the group of six cubes back again. "How much are we starting with?"

"Zero." Mac narrates the story I'm acting out. "Zero plus six, plus six, plus six,"

"makes...?"

"Um... eighteen! Okay, zero plus six, plus six, plus six, equals eighteen." Mac writes 0+6+6+6=18.

I ask Mac to look at 18-6-6-6=0 and 0+6+6+6=18 to see if he agrees that they tell opposite stories about eighteen.

"So they're kind of... So...." Max draws an arrow to connect the two number stories together.

"Can you think of a multiplication story using numbers and symbols that's the opposite of eighteen minus six minus six minus six equals zero?"

Mac looks at the addition equation he's just recorded. "Three groups of six equals eighteen!" He then writes 3x6=18. "So I'll write it down... Three minus... umm... times six... three times six equals eighteen." Finally, Mac really finishes the session by drawing an arrow from "this one" [0+6+6+6=18] to [3x6=18] and then to "this one also" [18-6-6-6=0]". He explains that the latter is "related because it's opposite"

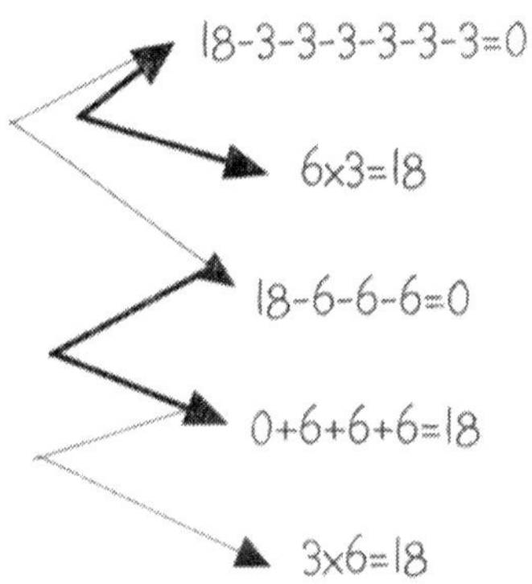

Mac's diagram showing the relationships between the equations

One of my intentions in working with Mac has been to enable him to develop different ways of seeing and expressing the mathematics he brings forth.... not through prespecified sequences of immersion but rather, by being attentive to and responding in each moment to the emergence of Mac's mathematics.

A second focus has been to make space for Mac to set his mathematics side by side — that is, put his ideas and expressive forms with one another and consider the patterns that exist among them. Mac's growth of understanding that arises from these spaces comes about through our interactions and his interplay of making numerical and operational sense of the game. Mac takes numbers and reconfigures them into equal groups through skip counting and physically rearranging cubes. He folds back[8] on what he now knows and extends his thinking to create yet other solutions by decomposing the number again or recombining the groupings. He reflectively examines his actions and develops not only compatible but complementary mathematical forms by making use of different operations.

At the end of the session, Mac creates another conceptualization that further integrates his understanding when he effectively brings all five equations together. Importantly, he's not linking one form to another, but comprehending them as interacting knowings of 18. By setting his mathematics side by side, Mac opens a space of his own where he relates and integrates the equations in ways that most definitely speak (mathematically) to one another.

Insight, I believe, refers to that depth of understanding that comes by setting

... to hear multiple points of views, as well as to express and clarify their own — is not seen as canceling the individual differences, but as a means of identifying them... within the context of the group [or, the larger environment][2]

experiences... side by side, learning by letting them speak to one another.[1]

Thinking systemically about students' mathematical learning raises the issue that as well as nurturing students' individual growth, emphasis should be placed on how their mathematics relates to larger collective realms.[3] Doing so allows students yet another opportunity to integrate different points of view and to gain an understanding for how mathematics exists simultaneously within individual and collective spheres of knowing.

Above is a piece of my thinking out on paper. Sometimes they are reflections about my teaching, my students' learning, something that I have read, or a combination of these. I call them "free ideas" because I write them very much in the moment as they arise. I do not pay attention to the particular words that I am using or the structure of the sentences. I am not even writing for anyone. I am simply letting my thinking out and paper is the medium that it happens to happen on. These ideas are also free in the sense that I write each one on a separate piece of paper and then put it in a basket with the others. Here, the intent is to use these ideas to prompt my thinking about specific issues. So as I continue to think out more, I sift through my ideas and consider how the one I have just written interacts with the others. Sometimes two or more ideas cluster together while other times, I purposely set seemingly contrasting ideas with one another to see if any generative friction results and enables me to think in a new or different manner. In both cases, I find myself asking — what does this mean? — and so it goes. Funnily, I am reminded of my childhood~ sitting at the table with my sisters and brother watching the green tea leaves swirl and sink to the bottom of our cups, creating images and each one of us trying to decipher the inherent meanings. Back to my previous idea... As I read what I have written out on this piece of paper, it makes sense and it has been integral in my current thinking and teaching. But I have not articulated it any further than this. So, here is the task I have set for myself. Two questions that emerge right away are: What does this fragment of thinking-out-on-paper mean for classroom mathematics and what might it look like in the classroom?

An example that immediately comes to mind and highlights the individual and the collective as necessary parts of the mathematics classroom is the lesson that evolved from the story, Even Steven and Odd Todd.[4] Had the lesson focused solely on the students' individual thinking and concluded with each child sharing and justifying his or her ideas, the students' understandings of odd and even numbers would have most likely existed like as my writings do when they are in the basket; as a collection of free and separate ideas. It is doubtful that the class, unprovoked, would have related their understandings to one another or created new ideas from making these connections. That said, had the lesson focused directly on establishing the class' group thinking, it is unlikely that the mathematical friction created by the different ideas which surfaced throughout the lesson and proved necessary for the students' further sense-making would have occurred. Rather, in scripting this unscripted lesson for the class, it was important that the space I opened was one that occasioned individual and collective mathematics.

With respect to individual knowings, several independently identifiable ideas arose in the lesson; all of which possessed some kind of rote or conceptual sense whether it was Mark's definition of even and odd numbers, Robby's declaration of odd numbers, or even Michelle's conjecture about adding two to a number. However, it was not until these ideas were brought together that the students formed larger collectives in which to work. By taking each idea and setting these into play, the class grappled with whether and how a certain notion contributed meaning to the concept of even and odd numbers. It was precisely in these moments as the students worked on their own and collaboratively with their peers that new understandings emerged and gave rise to ideas about even twos, partners, counting on by twos from zero, and splitting a number into equal halves. Moreover, it was the collective interaction of the ideas themselves that in turn extended the students' understandings and from which still new thinking came about: this time, ideas that were integrative in nature.

For example, Danica's conceptualization for why it is the last digit of a numeral that determines whether the number is even or odd can be seen as growing from the connection she made between the class' strategy of "even twos" and what she knew about place value. Further, what emerged from the students' examination of the entire set of proven ideas about even and odd numbers were relationships that connected all of these ideas together; through the notion of two, the students articulated exactly what was two about each idea and then distinguished the similarities and differences among the ideas as a whole. In these ways, the class' understanding cannot be viewed as having only been brought forth through the collective actions of the students but also, through the collectivity of their ideas.

Taking yet another reflective step back, the events of this lesson provide momentary glimpses into how the students' understandings arose through the simultaneity of individuals' actions and the individuality of ideas nested in and co-evolving with collectivity of ideas and collaborative ways of thinking. The collective qualities inherent in the students' subsequent understandings as individuals as well as that of a class were certainly much more than a set of accumulated ideas. At the beginning of the lesson, Mark's rules of even and odd numbers surfaced as isolated facts without any conceptual meaning. It was through the mathematical intra-actions and interactions or perhaps more accurately, the students' mathematical transactions that such meanings and relationships about even and odd numbers were created. These kinds of understandings are ones that could not have been achieved by any single child or as a result of piecing together the class' various ideas.

> Individual and collective understandings are not only necessary parts of the mathematics class(room), they are necessary for the growth of each other. (Co-)emergent in nature and co-existing, one enables the other.

Connecting as Shape-shifting

Connecting as shape-shifting? Hold on for a minute. What do you mean?

I suppose you could say that the mathematical thinking and emergent ideas that lead to a growth of understanding arise in the dynamic of learners simultaneously acting as individuals and as collectives.

So to use your favourite word, it's in the flow.

Exactly! And it's through which mathematical ideas and understandings shape-shift.

Okay... but show me an example of this shape-shifting of ideas and understandings.

Well, last week, the students had been given each a small bag of m&ms. During this activity, the students worked on their own: estimating and then finding out how many m&ms there were in the bag, sorting them by colour, graphing, and then determining the divisibility of the coloured sets by whole numbers.

Okay.

As I made my way to each of the children while they worked at their desks, I got a bird's eye view of the class' results.

You did, but your students didn't.

No. And they weren't looking and comparing their data with one another. Instead, they extended their initial analysis to general statements from what they had found in their own bag of m&ms.

Like...

Well, like James, for example, observed that for "all colours except green m&ms, there were two or three m&ms." And Holly explained that there were either "lots" of a certain colour or "hardly any" of it. Because these were small bags of candy, Holly considered "lots" to be four or more and one or two to be "hardly any."

So they were generalizing from their own particular instance.

Yes, and seeing the variations in the children's data, I wondered how I might bring their thinking to a collective realm so that the students could consider these conjectures within larger sets of data and other people's ideas.

So, you weren't just moving the children's thinking into this realm, but their ideas as well.

That's right.

But how would the larger context do more than create a collection of individual ideas?

Well, I was interested in what ideas might be born from the interaction.

... and how they'd shape-shift.

Exactly. So the next week, the children revisited their work...

Divided into groups of four or five, each child studied his or her bar graph,[5] wrote one statement that reflected a quantitative aspect of their bag of m&ms and then shared it with their small group. The bars of each graph read from left to right: dark brown, green, orange, yellow, blue, and red.

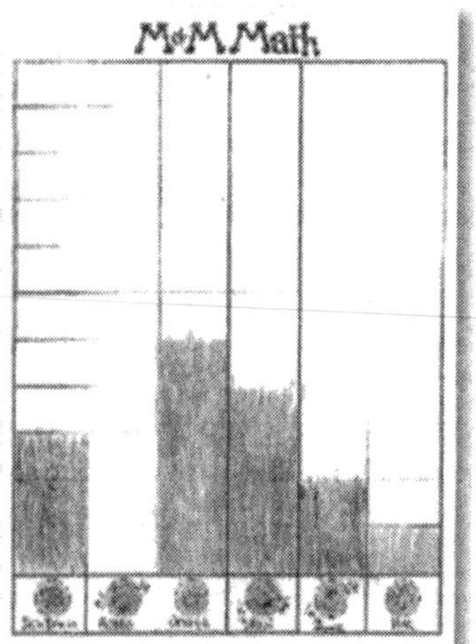

There is no same number of colors. (Robby)

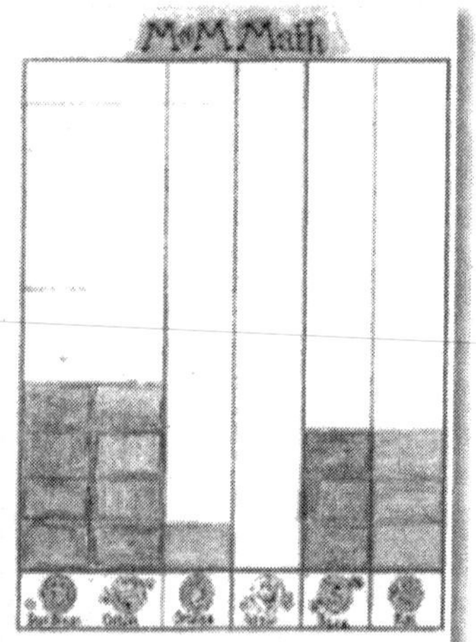

"There is Two pairs of the same numbers and another That's alone." (Michelle)

Green and brown don't have very many. (Mark)

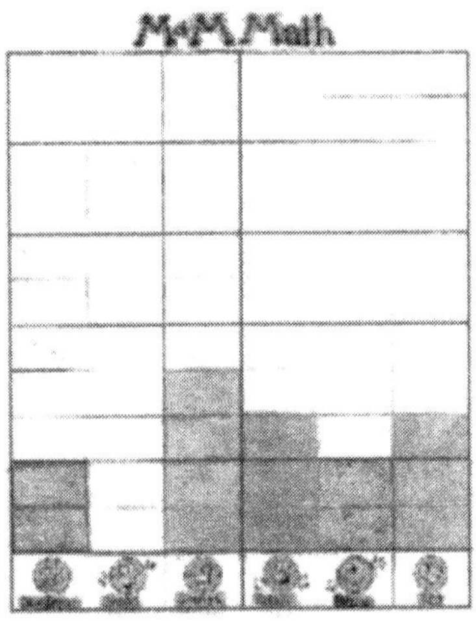

There is 3 reds and 3 yellows. (Danica)

"... another that's alone?"

I know! When I asked Michelle to tell me more about her statement, she explained that "because there's four here beside each other" and pointed to the four dark brown m&ms and the four green m&ms. "And three that are the same right there," pointing to the three blue m&ms and the three red m&ms. "... there's one that's all alone... it doesn't have a partner of a different colour," she concluded, pointing to the one orange m&m and then to the yellow m&m column of which there wasn't any.

She's using the metaphor of partner[6]... not to describe odd and even numbers but the similarities that existed within her sample. What happened next?

The students worked in pairs. Michelle with Danica and Mark with Robby. This time, each student took their "true" statement and examined it in relation to their partner's data.

To see whether or not it made sense for both sets of data.

That's right.

But what if a statement didn't reflect the larger sample?

They were to work out how they might revise it so that it made sense and still maintained its original idea.

Why didn't they just throw away the idea if it was incorrect?

Because I wouldn't let them! That wasn't the point of the lesson. It wasn't about right or wrong ideas but to focus on the notion of creating ideas in the face of other evidence so that the integrity of their original thinking could be maintained.

Oh, okay... focusing on how one's thinking and ideas can shape-shift.

Exactly. So when I went to join the two boys...

Mark compared Robby's graph against the statement – **green and brown don't have very many.** – that he had recorded on his sheet.

"How are you doing, Mark? Does it make sense?" I asked.

Mark replied, "sort of... because he has three brown and that's not very many and no green."

Having qualified three as "not very many," Mark appeared satisfied to keep his response the same.

I then went to see how Danica and Michelle were coming along with their work. While I did this, Mark picked up his pencil and underneath his first statement recorded:

green dosnt have any in Them.

Robby pointed to the sentence and commented that yes, he didn't have any green m&ms in his bag of candy but Mark did. When I returned, Mark told me that he had changed his statement to that of only green m&ms. As this did not reflect his and Robby's sample, I reiterated that they now to look at both graphs to see if the statement still made sense. Mark immediately shook his head and said no. I asked them how they might shift it so that it would reflect the larger sample of data.

"Keep the same answer," Mark said.

"So... keep *green and brown don't have very many*?" I asked.

Mark nodded his head and drew an arrow from his first statement down past the second one to show that he had changed his mind.

green and brown don't have very many.

green dosnt have any in Them.

In that episode, even though Robby explained very clearly, why Mark's second conclusion didn't fit for both sets of the data. Robby's attempt to shift his partner's thinking didn't have as much impact as I thought it might until you came back and began working with them. Why do you think this happened? What was it about your interjections that enabled Mark to reflect on his work and to see that his statement still made sense?

I suspect first of all, it's because they weren't used to working in this way — having to stay with an idea and consider how they thought about it even when it didn't immediately appear to resonate with the new context. Mark didn't think that his statement was true of Robby's data so he shifted his idea to only be about green m&ms; but when he did this, it no longer represented his data. Then, when Robby pointed out the contradiction, it might have been not as effective in prompting Mark to revise his statement because Robby moved from one set of data to the other. My interaction was slightly different in that I focused their attention on looking at the graphs together not separately, and asked them how the idea might be altered to accommodate both bags of m&ms. Not only recognizing the contradiction but also, the opportunity to make new sense within this context.

The boys examined Robby's statement:

There is no same number of colors.

in light of their data. This time, Robby moved his head back and forth on to discover that unlike his graph that had different amounts for every colour, Mark's graph had "... lots of the same... same number."

Interaction of the data occasioning change?

Yes.

Mark had two each of blue and red m&ms and one of each of brown & green. Mark suggested that they change the statement to, "... we have the same number of blue?"

Robby looked at the two graphs but did not say anything.

"Yes" I said, acknowledging Mark's suggestion but added "... then he's switching what his focus is because he's looking at *same amounts of colour*, right?"

Robby looked at his statement and then began thinking aloud, "...the amount of colour that...."

Mark and I turned to see what Robby was trying to work out.

"What do you mean?" I asked Robby.

"Do we have to combine them to get..." Robby asked, taking his pencil and pointing to both of the graphs.

"Yes. So you're talking about *same number of colours,* right?" I asked.

"Yes," said Robby.

Pointing to the statement on his piece of paper, I paraphrased what he and Mark had found. "You said that there's no same number of colours," I said pointing to Robby's graph. "But if you were to look at two bags of m&ms" I continued and pointed to the two graphs, "you'd find out that that wasn't so."

Robby shook his head in agreement.

"Because here," I pointed to Mark's graph, "Mark got some [i.e., m&ms] that had the same amount of green and brown and the same amount of blue and red. So, how could you change this idea so that it would make sense for both of the graphs?"

You helped them to define the

conceptual space they were working in...

The mathematical relationships of the context.

Yes and at the same time, pointed out the need for them to make new sense of that space.

Robby and Mark looked at their graphs. Mark decided that the statement could still maintain the notion of 'same amount' but rather than 'colour', he suggested that perhaps they could look at "same number in both [i.e., bags of m&ms]".

Mark began counting the total number of m&ms in Robby's sample, "three plus five is eight, plus four... is nine" he said as he added the dark brown, orange, and yellow m&ms. "... Is twelve, plus two is fourteen" he continued, adding the number of blue m&ms. "Fifteen" he concluded after adding on the one red m&m.

Robby also found the total number of m&ms in his sample but did the calculation in his head, nodding while Mark subtotaled aloud and giving one final nod when he reached fifteen.

Mark then totaled his number of m&ms, "... and then you'd go two, four, six, eight, ten, twelve..." while counting the dark brown, green, and orange, and yellow m&ms by twos and adding the four remaining blue and red m&ms and ended with "... sixteen."

"Okay, he's got fifteen and you've got sixteen," I said and looked back and forth at Robby and Mark. "So what could you do with that idea [i.e., 'same amount']? Could you say something about the number of m&ms that would make sense for both of those?" I said as I pointed to the graphs.

"Maybe, not all bags of m&ms have the same number in them." replied Mark.

"What do you think, Robby?" I asked.

Robby nodded his head, "yes." For his second statement, he wrote:

Not all m and m bags have the same amount of m and ms.

You can see how the students' thinking progressed in the way they developed the statement. Staying with the notion of 'same amount' created the need for Mark and Robby to re-analyze their data and form a related yet different interpretation.

And the result wasn't the effort of just one individual. What came about was a consequence of mathematical interaction amongst the three of us, all thinking around the idea of 'same amount'.

What were Danica and Michelle doing?

The girls had taken Michelle's idea and tested it against both of their graphs. They were smiling and calling me over to see what they had found out.

"How'd you do?" I asked.

"Good!" Danica said. She picked up her graph and showed that "because I have the same as hers... like two, two," Michelle's statement remained true. Danica had the same number of red and yellow m&ms and the same number of blue and dark brown m&ms in her bag.

Michelle supported Danica's explanation by reiterating that "we have two pairs that are the same" and also added, "and then we both have one that's different." This latter statement was in reference to what Michelle had identified as "one that's all alone," which she described as "one that's all alone... it doesn't have a partner of a different colour."

Like Mark, Michelle drew an arrow underneath the first statement to indicate that the statement remained the same when compared against both sets of data. And just as in Robby and Mark's case, Danica and Michelle were also challenged when they came upon examining their graphs in relation to Danica's statement.

Danica commented that, "Michelle has three reds, no yellows, and three blues." She then wrote:

Michelle has 3 reds but insted of 3 yellows she has 3 blues.

Both girls agreed that this only made sense for Michelle's graph and that it did not apply to both of the graphs.

So then, what?

Well, just like Robby and Mark, they could say what it wasn't but they weren't sure how to shape-shift the idea.

Bringing forth Danica's notion of the number of red and yellow m&ms, I repeated what she had said but focused their attention on the red and yellow m&ms. "And you found out that Michelle has no yellows but the same amount of reds... could you do something... with the idea — looking at the idea of reds and yellows? What could you say? Both graphs..."

"... have reds in them.... Both graphs have three reds." Danica replied.

Michelle and Danica double-checked the statement to make sure that it reflected both sets of data. The two girls nodded and concluded "yes!" Danica then recorded:

both gragphs have 3 reds.

Having examined and revised their ideas against their partner's data, I asked the children to form a working group of four.

I read out the statements that they were to now focus on:

- **green and brown don't have very many.**
- **Not all m and m bags have the same amount of m and ms.**
- **both gragphs have 3 reds.**
- **There is Two pairs of the same numbers and another That's alone.**

As a small group, we reflected on the statements and talked about how two of them had undergone changes while the other two remained the same. The group then began to test the statements to see whether they satisfied all four sets of data. The children laid their graphs out in a row and examined them, beginning with Michelle's claim:

There is Two pairs of the same numbers and another That's alone.

Mark quickly pointed out that the statement no longer made sense because it did not reflect his or Robby's graphs. The four children studied the graphs, saying nothing. To help them, I suggested that they might make sense of three of the graphs rather than all four at once.

Robby looked at the girls' graphs, then Mark's, and then his. Here, he corrected Mark by pointing out that Mark's graph was coherent with Michelle's pattern because it too had two pairs of colour sets with the same number of m&ms; two of each blue and red m&ms and one of each of dark brown and green m&ms. Robby swept his hand across his graph and explained that it is when they come to his data that the claim does not hold true.

"Mine is the hardest because it doesn't have any pairs. So, it has nothing to do with that," Robby said, pointing to Michelle's statement.

Looking from a different perspective, Danica observed that Mark's, Robby's, and her bag of m&ms all had two blue candies. However, she dismissed it when she realized that Michelle's bag of m&m had three blue m&ms.

"Each of us has four of... well, not of the same colour... but four of *whatever* m&ms," Michelle said.

"Four of *any* colour" said Robby.

"Yes." Michelle said.

"Everybody has four of every colour..." Danica concluded.

Michelle mulled this over. "Four of...."

"Four of *any* colour," Robby said again, correcting the language of the conclusion so that it would be true for all of the samples.

The group agreed with this and made the conjecture that a bag of m&ms would have four m&ms of some colour. Michelle was not sure how certain they really were that every bag of m&ms would contain four candies of a particular colour and hesitated in committing this to paper. She and the other three children decided that they were not completely certain and that the term, "most" needed to be included in their statement. Michelle picked up her pencil and wrote:

~~Any~~ Most bags have four of the same colours.

In that episode, you weren't directly interacting with the students. You were there, but the children worked on their own... on their own, as a collective.

Their thinking and work is an example of what I'd consider to be a collective form of shape-shifting.

I think I know what you mean now. Each time a new aspect was proposed, it was examined and assessed by the children in terms of whether or not it helped to further define the idea before it was integrated with what they had. Mathematical thinking and mathematics in this collective, shape-shifting manner, is really, a recursive and co-evolving process... emerging thoughts, melding or not with that which already exists and through revision, there's renewal of thinking, of ideas...

The students moved on to Danica's idea of three reds. Scanning the four graphs, Danica shook her head and answered, "no." Neither Mark nor Robby had three red candies in their bags of m&ms. Danica adjusted her statement in response to the new data and announced that, "everyone has reds."

Robby, Mark, and Michelle all nodded and agreed with an enthusiastic "yes!"

Still keeping the idea of red m&ms...

But no longer on a specific quantity...

The new statement now satisfies all four samples.

Danica crafted it a bit more and recorded:

every body has atleast I red.

Robby read out his statement, "**Not all m and m bags have the same amount of m and ms.**"

Mark, Robby, and Danica skip counted by a combination of twos, ones, and threes and found out that Danica had fourteen m&ms and Michelle had fifteen candies in total. The boys no longer thought that the statement made sense.

Robby justified this, "because Michelle and mine have the same amount."

I reread Robby's statement, emphasizing the language in it, "It says that *not all* bags of m&ms have the same amount. Does that mean that *some* bags *could* have the same amount?"

"Yes," said Robby and Mark.

Robby then drew a downward arrow from his second statement to show that the third statement remained the same.

Not all m and m bags have the same amount of m and ms.

As he did this, I prompted the other children to think about the meaning of the statement and to articulate why the use of the word, "not" made it work. "It wouldn't make sense if Robby said... what?"

"*All* the bags of m&ms have the same amount in them." said Mark.

With this, the group moved on to testing the fourth and final conjecture that was Mark's claim that **green and brown don't have very many.** Immediately, the group saw that it held true for Mark, Robby, and Danica's graphs but not for Michelle's. She had four of each of green and brown candies and they agreed that given the small package of m&ms, four was too many to be considered "not very many."

Wanting to integrate the language used in Robby's statement, Mark thought aloud, "not all bags of m&ms..."

Danica interjected, "all bags don't have the equal amount of browns in each column. Like... nobody has the same amount in browns 'cause Mark has one, I have two, Michelle has four, and Robby has three."

Mark agreed with what Danica said and was about to write: "not all the bags have the same number of brown." when Robby spoke up.

"I have another one that he might want to use," he said. "It is... not all bags have very many... not all bags have very many greens."

"But when you look at Michelle's, she has four," said Mark.

"I said *not all,*" responded Robby.

"So...?" I said.

Mark paused for a moment. "I would say that not all bags have the same number of green and brown in them!" He recorded the revised conclusion:

not all m and m bags have the same number of green and brown.

Another recursion in his thinking that effectively integrated the notions of "not all" and "sameness" together while still keeping the focus on green and brown m&ms.

Prompting one final analysis, I gave the children two more sets of data to consider:

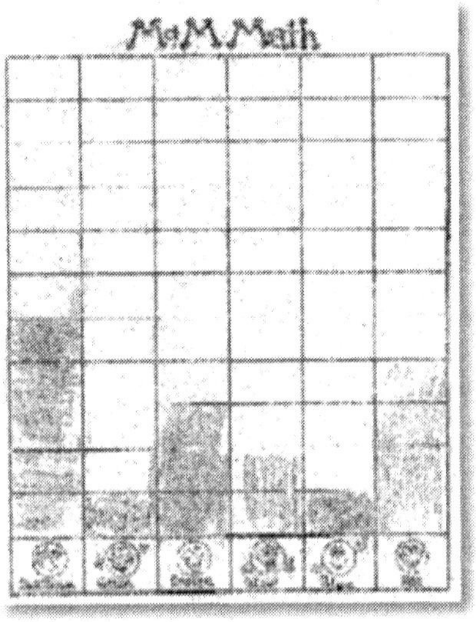

Lara's graph

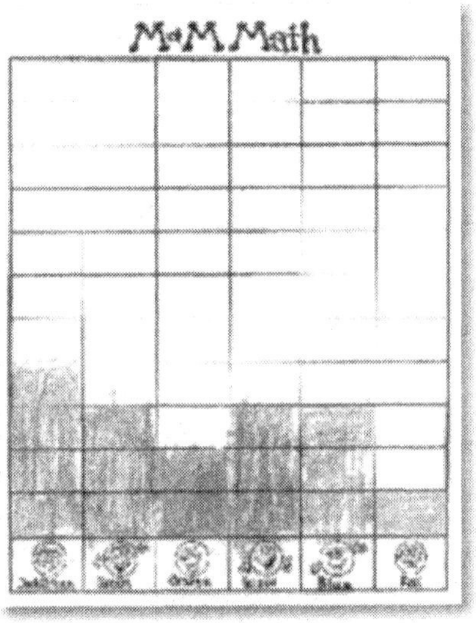

Amanda's graph

Now with six graphs to consider, the group found that Michelle's statement remained the same. As Lara had four dark brown candies and Amanda had five dark brown and four red m&ms, both samples satisfied the conjecture that most bags would contain four m&ms of a particular colour.

Danica smiled as she nodded to indicate that her statement was also satisfactory. According to the two new graphs, Amanda and Lara had at least one red m&m.

Robby read his statement aloud, "Not all m and m bags have the same amount of m and ms." Following this, he totaled the number of m&ms for each of the two samples in his head. Robby concluded Lara and Amanda's bags each contained sixteen m&ms.

"Does that hold true still?" I asked.

"Yeah, a bit" he said.

"Why are you saying *a bit* now. What have you found?!" I asked.

"Fifteen goes with mine and Michelle's and... those ones," he explained as he pointed to Lara and Amanda's graphs, "goes with Mark's".

"So would you want to change your idea or shape[-shift] it a little bit?" I asked.

"Yes. Most of the bags have fifteen or sixteen" he said and then recorded this as his final statement.

Do you see that? Do you see what's emerging from Robby's thinking?

Yes! He's looking at a different, more specific pattern in the data. He didn't identify it as average number of m&ms per bag, but that's the distinction he's making — that of the bags of m&ms, more of them had a total of either fifteen or sixteen candies than not.

Mark read out his statement and quickly nodded "yes" — that it still made sense for the six sets of data. According to the graphs, both Amanda and Lara's bags of m&ms had different quantities of green and brown when compared with the other four graphs.

Having tested and revised their conjectures across the six samples, the children told me that although they were "pretty sure," they were not completely certain that the four conclusions would hold true for even larger samples of m&ms.

I wonder what would make them very certain of their conjectures. Did you ask them?

Yes, and...

Robby figured that "you'd have to look at as many bags as possible twenty to forty" while Michelle was still not convinced that this would be enough samples to be really sure of her statement.

She said that she would need to look at many more — "about twenty-five bags every day for twenty-five days!"

Danica agreed with Robby and explained that a sample of twenty bags would suffice in testing whether every bag of m&ms had at least one red candy. And Mark said that he would need to test between fifty and seventy-five bags of m&ms to be certain of his conjecture.

Bringing this session to a close, I asked the children to look at their statements and to consider if and how they changed as the number of samples increased.

"Some of them changed and some stayed the same," said Mark.

"Did anyone have an idea that stayed the same for the whole lot?" I asked.

All four students shook their heads "no."

Having established that all of their conjectures went through some kind of revision, I was interested in their thinking on why this happened.

"So everyone's changed at some point. Okay. Why did they change?" I asked.

The group explained that each time they looked at the results of another bag of m&ms, they realized that their conjectures were not always consistent with the new data "because not all m&m bags are the same."

Mark, Michelle, and Danica all noticed subtle shifts in the shape of their conjectures over the course of examining the six samples whereas Robby nodded his head and exclaimed that his work underwent "*HUGE* changes. I started with colours of brown and green and then with all colours [i.e., total number of m&ms in a bag]."

By taking the mathematical patterns that the students identified from each of their graphs and situating them within larger, collective realms Mark, Danica, Michelle, and Robby got firsthand experience in how the mathematical ideas shape-shifted... and how they evolved as a result of the interaction of their individual and group thinking.

And, the integrity of the final conclusions that were born from these interactions... they were really, much more than the sum of the parts that fed into them. I mean, the specific aspects generated by the children as individual and collective agents came together not in a piecemeal, jigsaw puzzle way but in a co-emergent manner. What they gave rise to were very sophisticated ways of thinking about the data.

More specifically and pervasive in all of the conclusions was the understanding of what you can generalize from the data and what you cannot. As well, that growth of understanding isn't always a matter of discarding and replacing ideas but rather, seeking out patterns... relationships.... and through shape-shifting... deeper, more comprehensive notions within a concept can be developed — developing thinking into a greater whole.

Striking a balance between those spaces in which the students and I are involved in direct mathematical interactions and spaces where I am not part of their immediate work is another important aspect of place-making in the classroom. When I am working with a child or group, I am in the midst of, and therefore, part of the co-emergence of mathematics taking place. On the other hand, when I am not directly interacting with the students I become part of the extended environment and opportunities of a different kind arise for the students and myself.

For the students, they are responsible for opening and exploring their own mathematical spaces. For me, I am able to attend to and learn about their understandings from a more distant vantage point. The spaces that are created as a result of the students' mathematical intra or inter-actions not only provide me with glimpses into their understandings but also, sources from which to further investigate their thinking, perhaps by engaging them to fold back[1] or provoking them to extend their mathematics through direct interaction.

Opening Mathematical Spaces Of Their Own

With a blue pencil crayon in hand and a piece of paper to record their mathematics, Sammy makes his way to the round table and pulls up a chair for his best friend, Sam, so that they can work together. Sam sits down and looks around the classroom.

"Sam," Sammy taps Sam on the shoulder. "Let's do it!" Sammy asks Sam what he knows about the number, 72.

"How about..."

"What equals that number?!" Sammy raises his right hand in the air. "Oh! I Know!"

"How 'bout... divided!" Sam leans towards Sammy and clasping his hands together smiles, "... will make it equal..."

"How about times? We could say something that equals that number."

"Yeah!"

"What is it? Don't say anything! Just let me think." Resting his chin in his hand and his elbow on the table, Sammy looks in the opposite direction from Sam. Sammy then records 8x9=72, which catches Sam by surprise.

"How do you Know that?"

"I Know the times-table!"

Sam and Sammy are searching for something they already know to be true about the number, 72. They don't intend to develop a mathematical description for the number but to re-member or fold back and collect an image[2] — to search through their mathematical understandings and locate something they know, without question, to be a fact about the number. Although Sammy and Sam specify that they want a multiplication fact, they aren't concerned about locating a specific one. Sammy's image of 8x9=72 satisfies this because he simply retrieves it. He hasn't had to engage in any mathematical activity to come up with this fact he just knows this to be true because of "the times-table!"

"My turn! Seventy-two plus seventy-two..."

"That number plus that number divided by two equals that number.... Two, two..." Sammy then proceeds to add the two sets of twos in 72 + 72 together.

"It's one hundred forty-four! So it's one-hundred forty-four divided by two!"

Sam pleads with Sammy. "Let me count... Seventy..."

"Multiplied by two."

"... eighty, ninety, one hundred...." Sam counts on, first aloud and then whispering to himself, keeping track of the number of tens he is adding by unfolding a finger each time. Once done, he then writes 144÷2=72.

What's going on here? Sam's announces that it's his turn to describe 72 and the boys return to Sam's first idea to include "divided!". They set to work on developing a mathematical expression that uses this operation and now, different aspects of their understanding are revealed. Sam on the one hand, approaches the task by using repeated addition to find the sum of 72 to 72 and records 144÷2=72. From his working with specific numbers and his insistence that Sammy let him skip count on his own to determine what the sum for 72 + 72 is, I assume that Sam is using what images he has to come up with the mathematical expression. He doesn't appear to know this already and his actions indicate that he doesn't have a more formal understanding of the pattern he's working with… like, 'if you add a number to itself or double it and then divide it by two, you'll end up with the original number.'

Now Sammy on the other hand, tries to complete Sam's thinking by explaining that, "that number plus that number divided by two equals that number". His thinking is a more formal conceptualization. Where Sam works with specific numbers, Sammy's considers the mathematics from a more distant perspective. He's thinking about how the actions of repeated addition and dividing by two, work for any number and then he's applying this understanding to the specific context of 72. As well, Sammy's suggestion that Sam "multiply by two" implies that he understands repeated addition to be related to multiplication.

Pause.
Giggles.
Pause.

Sam suggests taking 72 away from a ridiculously large number.

More giggles.
Pause.

"Aa-she-waa!" Sammy rests his head on his hand.

"Oh! Let's say something about it. Let's not do any of the sums."

"It is between eight and something," Sam extends his arms outwards and moves his body in a left to right motion.

Sam's gestures tell me that he's thinking with an image of numbers organized along a horizontal line. He expresses 72 as being located between eight and another supposedly, greater number.

"It's between seventy and eighty." Sammy records: It's between seventy and eighty.

"That's a good one." Sam smiles at Sammy.

"Is it odd or even?!" whispers Sammy.

Sam takes the pencil and writes: It is a even number. and then states, "It's a two digit number."

"It's a two digit number." Sammy, smiles back at Sam.

Sam writes: It is a two digit number.

Sammy's prompt for them not to just continue to perform operations on 72 but offer other ways of thinking about 72 reveals that these two students associate numbers as possessing other properties rather than only being a product of arithmetic actions.

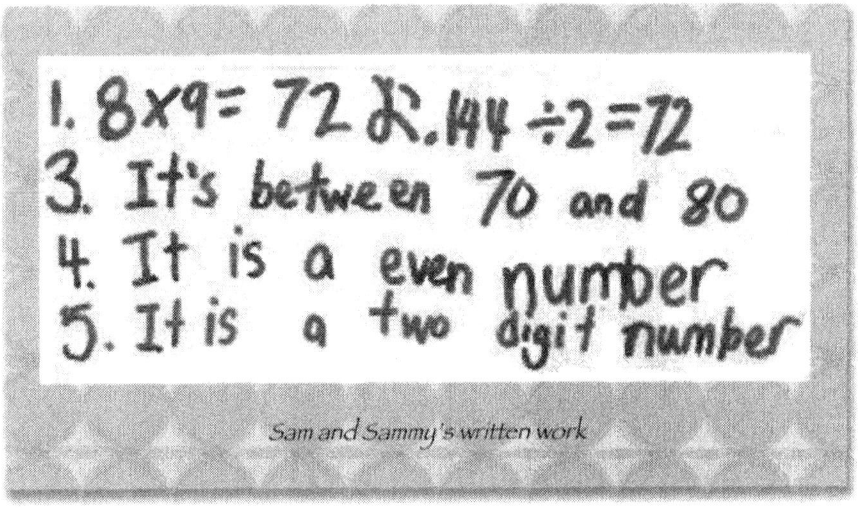

Sam and Sammy's written work

I approach the two boys. *"Sam and Sammy, may I join you?"*

Sammy moves the piece of paper between the three of us and begins reading the equations that they have written about 72. He tells me what he has already told Sam — that he knows 8 x 9 to be 72 because he knows his times table.

This confirms my previous assessment that he's simply folded back and collected an image. Now, I want to know more about the thinking that gave rise to what Sam has written for #2.

"How did you think of doing one hundred forty-four divided by two equals seventy-two?"

"Because seventy-two plus seventy-two equals one hundred and forty-four. If we divided by two, we'd be cutting one hundred and forty-four in half."

In this one sentence, Sam reveals many aspects about the understandings that are embodied in the mathematical expression. Not only is the equation numerically and operationally correct, but he's also able to step back and talk reflectively about the relationships and distinctions he's made about the mathematics at hand — property noticing.[3] He considers repeated addition and possibly, multiplication by "doubling" 72 as being the opposite or inverse action of division. He relates "cutting one hundred and forty-four in half" as being the same as dividing the number by two. Because Sam is focusing on specific values — 72, 144, and 2, it's not possible to say whether he's got a generalized understanding for what happens when the combination of operations are applied to any or all numbers.

I ask Sam and Sammy if they've recorded any descriptions other than symbols.

"Yeah" Sam points to #3 on their sheet. "It's between 70 and 80".

"Yeah. We used words."

"Can we move on to that and look at that?"

Sammy reads aloud, "It is between seventy and eighty."

I ask Sam and Sammy why this makes sense to them.

"Because it is."

"Because it's higher than seventy and lower than eighty."

Sammy's image of 72 as "higher" than 70 and "lower" than 80 are similar to Sam's previous bodily gestures. He too appears to visualize a number line. Sammy's number line however, is a vertical one on which 72 is located "between" 70 and 80 and "ranked" higher than 70 and lower than 80. But where did this image come from? Is it simply a linguistic one that he learned by rote? Or is it an expression of conceptual number sense? I suspect that it's connected to the game we play in class where the children try to make the highest (or lowest) possible number by placing randomly drawn digits — 0 through 9 in either the hundreds', tens', or ones' column. In this game, the class creates a vertical number line by recording all of the possible numbers that could have been made with the drawn digits on the chalkboard. This part of the game requires them to identify and describe to their peers whether and sometimes, why the number is "higher" or "lower" than the other numbers already recorded.

Moving on, I read out their next description — "It is an even number" and ask the boys to talk about it.

"Because if it [72] was a seventy-three, it would be odd because the seven and the three are odd....". Sammy in a more confident tone of voice, "It's [72] even because seventy-two can be divided by nine... it can be... divided into nine, into nine groups." Sammy then points to 8 × 9 = 72 on their paper.

"Okay, so you're saying if seventy-two is even, it can be divided..."

"Because eight times nine means eight groups of nine, equals seventy-two."

"Okay..." I ask the boys what they think of 73. They immediately in unison, tell me that 73 is an odd number.

"It would[n't] be divided into equal groups.... Because it can't be divided by nine, it can't be divided into nine groups."

I'm interested to see whether Sammy will elaborate on this notion of equal groups by applying his thinking to other numbers.

I pose the question, "what if it [the number] was seventy-one or seventy-four?".

No response! "What makes an even number... even?!"

"You can divide it into equal groups."

"You can count by twos. Two's an even number [unfolds two of his fingers], four [unfolds four of his fingers], six [lays his right hand over his left hand so as to make 'more' fingers], ... all the even numbers."

Based on the fact that Sam recorded 72 as an even number and Sammy didn't disagree, I might have assumed that they shared the same understanding for why 72 is an even number. However, now that they've revealed what they each mean by "even", it's clear that Sammy and Sam have two completely different understandings.

Sammy understands "even" numbers to be those numbers that can be "divided into equal groups". In this case, 72 is considered to be an even number because it can be shared evenly into eight groups of nine without any remainders. Of course, this is true of even numbers, but it's not this quality that defines them as such. Sammy also explains 73 as being an odd number that "can't be divided into groups of nine". Because 73 is both an odd number and a prime number, and as well, its digits — 7 and 3 are odd and prime, it's difficult to distinguish what he means by "odd".[4]

In contrast, Sam's thinking of even numbers comes from a mathematically different place of knowing than Sammy's. Sam's understanding of what makes a number even or not is in fact, the appropriate one. As he demonstrates for Sammy and me, a number can be distinguished as being even (or not) by whether you can 'arrive' at the number by skip counting by twos.

[reading #5] "It's a two digit number. What do you mean by that?"

"Because there's two numbers."

"It's got tens and ones."

"Okay, what's a ten and what's a one?"

"There's seven tens and..."

"... and there's two ones."

"What does seven tens mean?"

Sam explains that if you take the base ten blocks and use five ten rods and one unit cube, the resulting number is 51. He draws this with his finger.

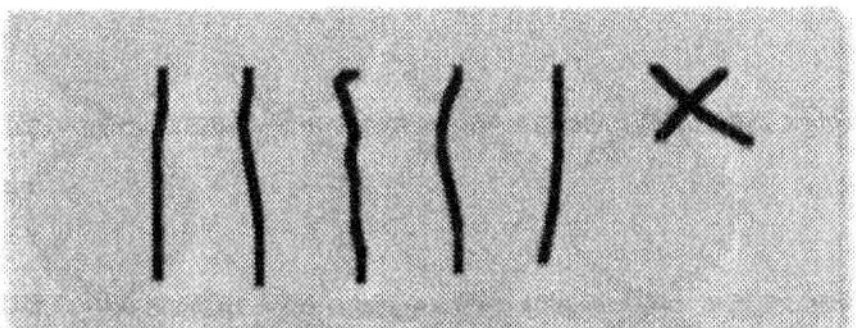

Sam's drawing

Sammy and Sam explain to me that seven tens can also be thought of as, "seven groups of ten" or "seven times ten."

The boys demonstrate that they know numbers can be identified as having "digits" and in the case of 72, the digits "7" and the "2" make it a number consisting of a total of two digits. At the same time, they also communicate their understanding of place value. According to Sammy, 72 can be considered as having "tens and ones". Sam points out that 72 has "seven tens" and Sammy finishes by explaining that it also has "two ones". When Sam acts out that 5 "tens" blocks and 1 "unit" block would be 51, he reveals the understanding that the first digit of a two digit number refers to the groups of ten in that number and the second digit refers to the number of ones or units it has. The boys' further discussion of how seven tens can also mean "seven groups of ten" or "seven times ten" tells me that they know the left-hand digit in the two digit number as having a value that is ten times greater than the digit to its right.

I ask Sam and Sammy if there is anything else that they know about 72. I leave the table so that they can work on their own again.

"Let's just draw one more. Groups. Groups

of what?... I Know! Let's draw a group story and then..."

Sammy grabs the pencil from Sam's hand and moves the paper in front of him. "Let's draw... Let's draw... There's seven tens, right? We can draw seven tens."

Sam pauses. "Oh... we draw ten groups of seven and one group of two."

"SEVEN groups of ten!"

"TEN groups of seven!"

"It's *backwards*."

"Put ten circles, with each...." Sam draws the diagram with his finger to try and explain his thinking to Sammy.

"It's seven strokes for seven tens, remember?"

"Oh yeah!!"

"One, two, three, four, five, six, seven." Sammy draws and counts out seven large circles.

"One, two, three, four, five, six, seven, eight, nine, ten." Sam draws ten Xs in the first circle.

Sammy's persistence that Sam's image of ten groups of seven and two ones is backwards and that they should record a diagram of seven tens and two ones suggests that Sammy's thinking is firmly grounded in its place-value meaning and thus, doesn't consider the former as being an equally valid way to describe 72. Sam's flexibility to move from one image to the other however, indicates that he does relate the two images as meaning 72.

The two children continue taking turns at drawing the diagram until all seven circles are filled with ten Xs each and the last circle has two Xs in it.

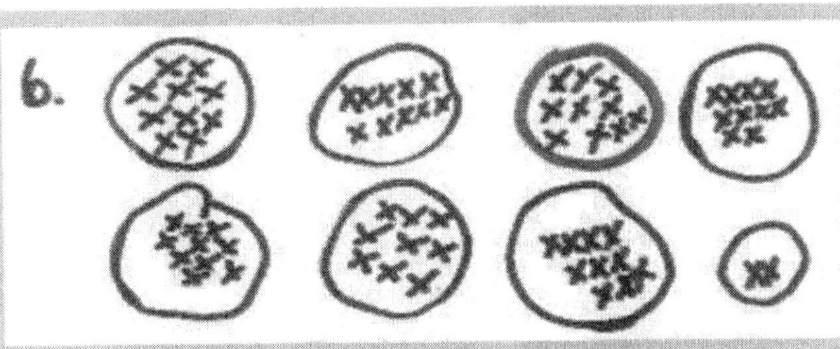

Sam and Sammy's drawing of seven groups of ten and two ones

I return to Sammy and Sam's table. Sammy is telling Sam that he has thought of another way to describe 72.

Just when Sammy explains that seven tens and two ones make 72,

Sam exclaims, "groups!" and writes 7 groups of 10 and 1 group of 2 makes this number.

7. 7 groups of 10 and 1 group of 2 makes this number

Sam's clue in response to Sammy's description of their previous drawing

I'm curious how they might express their thinking with symbols and so I ask Sam and Sammy if they can think of another way to describe it.

Sammy and Sam talk aloud as Sammy records 10+10+10+10+10+10+10+2=72 onto the paper. "Ten plus ten plus ten plus ten plus ten plus... one, two, three, four, five... plus ten plus ten equals 72."

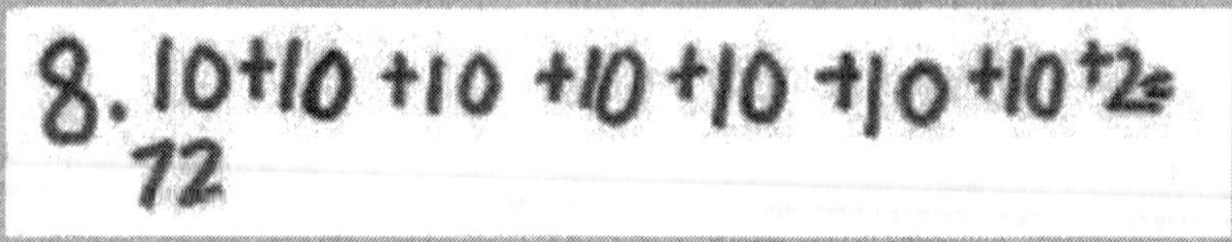

Sammy and Sam's eighth clue

The students' work demonstrates their fluency in moving from one mathematical form of language to another — spoken, written, diagrammatic, and symbolic.

Sammy offers another suggestion. "How about minus?"

"How about repeated subtraction?!"

"Oh. I think I know one." Sammy points to 144÷2=72 and says, "How about one hundred forty-five minus seventy-three equals seventy-two? Just make this — one higher" as he points to 144.

"Maybe we'll make a repeated subtraction."

"How will that work?"

"One hundred forty-four minus ten..." Sam begins to record, 144-10.

"Oh! I know!"

"... minus ten, minus ten..."

"You need four." Sammy keeps track of the number of tens Sam has subtracted from 144.

"minus ten"

"five"

"minus ten"

"six"

"minus ten"

"seven"

"minus two"

"equals..."

Sam finishes the equation.
144-10-10-10-10-10-10-10-2=72.

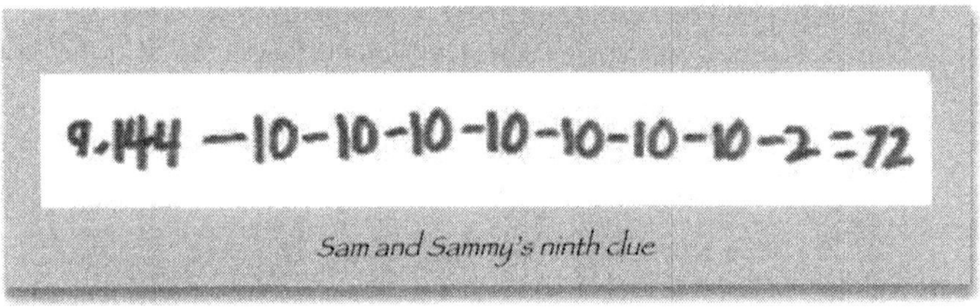

Sam and Sammy's ninth clue

I point to clue #9. "How'd you come up with this?"

Sam points to,
10+10+10+10+10+10+10+2=72 and tells me "I can read the plus."

Sammy adds: "Forty-two (sic) [presumably, 72] is multiplied twice — two seventy-twos." He points to the 72 in the equation, 144÷2=72 and explains, "And if you minus one of the, take away one of the seventy-twos, but actually don't say something [i.e., 72], do it another way, like this one," at which point he sweeps his finger over 10+10+10+10+10+10+10+2=72, "you get seventy-two.

Equals another seventy-two. Sammy points to =72 in 144 -10-10-10-10-10-10-10-2=72.

By asking Sammy to explain his thinking reveals yet other sophisticated aspects of his mathematical understanding — once again, qualities that aren't visible by only looking at what's been written. Sammy explains that instead of writing 144-72=72, he wrote, 144-10-10-10-10-10-10-10-2=72. The reason why he didn't record it as 144-(10+10+10+10+10+10+10+2)=72 is simply because he hasn't learned how to use brackets yet. Regardless of this, it doesn't stop him from thinking conceptually about the mathematics in this manner. Together with his physical finger pointing to 10+10+10+10+10+10+10+2=72 and verbal reference to "take away ONE of the seventy-twos but don't actually say [it], do it another way, like this one" illustrates that Sammy's very much 'standing back' and reflecting on or observing his formal understanding of the mathematics. He's not speaking about the repeated subtraction of ten but is using a specific case of the distributive law over addition:

$$-(10+10+10+10+10+10+10+2) = -10-10-10-10-10-10-10-2$$

Even though Sammy hasn't generated a 'theorem' as such, he is in fact, using a formalized understanding of addition in this context.

I want to know what Sam's thinking and so, I don't react to Sammy's explanation but continue to listen.

"Seventy-two plus seventy-two equals one hundred and forty-four." Sam does this while he stretches his thumb and fingers around either end of the, -10-10-10-10-10-10-10-2 and says, "Seventy-two."

"Sam just took another seventy-two by not saying it."

I repeat what Sammy just said but in a louder voice to try and bring Sam into the conversation. "Sam just took another seventy-two by not saying it?"

"By not... writing it down." Sammy sweeps his fingers in a circle around, 144-10-10-10-10-10-10-10-2=72.

"By not... like... one hundred and forty-four minus... you minus one hundred and forty-four by seventy-two." Sam also sweeps his finger across 144-10-10-10-10-10-10-10-2=72, "and you get seventy-two. Because... because..." and points to 144÷2=72.

"But you also minussed! ... Sam... Sam also minussed seventy-two but he also... one hundred forty-four minus seventy-two equals seventy-two. He also minussed the seventy-two."

By encouraging them to continue, I see that Sam's also thinking of 72 in the same manner as Sammy. Although the boys don't talk in general terms but repeatedly use clues #8 and #9 as specific examples in their explanation of how they "minussed seventy-two", neither of them speak of "144 minus 10, minus 10, minus ten." and so on. It's clear to me that BOTH students think of subtracting 72 in the form, 10+10+10+10+10+10+10+2 as they have expressed in #8.

A while later, I ask the Sam and Sammy to revisit their work. They return to the eighth clue and begin working with other values — 85, 37, 1010, and 9. And just as they had done previously with 72, they represented each number "by not saying it" and showing "it another way" by creating an addition equation.

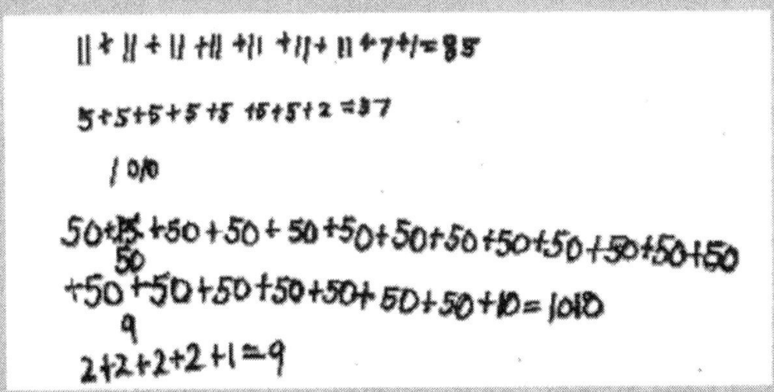

Additional examples of Sammy and Sam's "rule" that they used in clue #8

Moving to their ninth clue, I ask them if what they've recorded here could be true for any number.

In other words, can they extend their thinking to a more general understanding of this concept?

"I'd say yes, most of them..."

"Most probably." Sam turns to Sammy. "Pick a number. Pick a number and I'll try."

"Okay. Try the first number we did", referring to an equation ||+||+||+||+||+||+||+7+|=85 that they had written earlier. "Eighty-five and..."

"Just pick any number. Just pick a number."

Sam and Sammy now take what they know and apply it to a new context. I'm interested in how they might use the mathematics as a pattern and not, in a situation-specific manner.

Interrupting them, I shift the conversation by bringing Sam and Sammy back to their previous conjecture. **"Sammy, you said that you thought it would work for any number because it was almost the opposite. What did you mean by that?"**

"It's subtracting instead of adding." Sammy refers me back to what he had recorded as, ||+||+||+||+||+||+||+7+|=85.

"... to do this... maybe the number you pick, you add it, you add it with the same number that you... picked... you get the big number..."

"You times it by two! Minus it"

I'm assuming that by "it", he means the number, 72.

"Like eighty-five plus eighty-five equals one hundred and seventy... and you minus it." Sam points to 11+11+11+11+11+11+11+7+1.

"To get rid of the extra half."

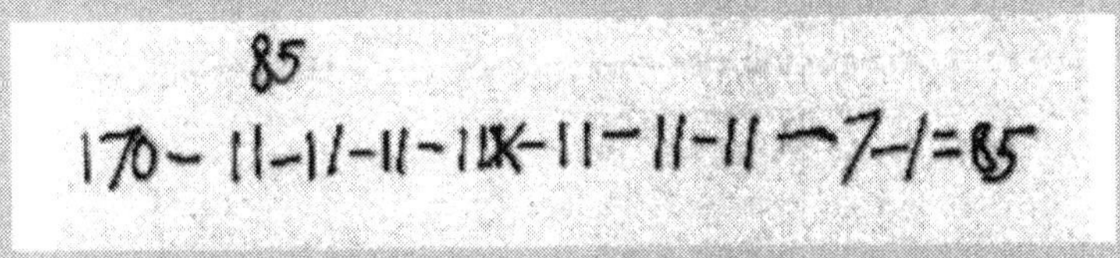

Sam and Sammy's example of how their "rule" from clue #9 of 'add the number by itself/multiply the number by two and then minus it by "not saying it"' applies to other values such as, 85

After listening to Sam and Sammy explain their thinking again but this time in more general terms, they definitely demonstrate that what they know isn't specific to particular numbers but indeed, a formalized pattern that they've reflected on.

Watching from the edge and occasionally jumping in, I was able to observe Sam and Sammy as they opened up and explored their own mathematical space. Their work was not concerned with generating as many arithmetic facts that they could think of but instead, thinking creatively about 72. Sam and Sammy's activity in folding back enabled them to thicken and extend their mathematical ways in which to express 72. For me, Sam and Sammy's mathematical space served as an opportunity to view the complex and integrated understandings that these two students have developed about natural numbers.

Mathematics Beyond the Classroom

I have been working to define another direction from which to come at G. Bateson's concept of, patterns which connect. I want to teach collaboratively with another colleague who shares a similar interest and vision for teaching mathematics. As well, because my students will already be working with other students who are younger and older than they are and, there is not another second and third grade class in my school, it would be great if the teacher also taught children of the same ages. I want the collaboration to be an active, ongoing engagement for the teachers and students. I see this as an opportunity not only for sustained cooperative learning where long term relationships can develop for the children's mathematics to take root, connect, and grow but also, an occasion for co-emergent and collaborative teaching.

Two weeks later...

How can I make this happen? More specifically, with whom?! It has to be with a colleague who not only has a passion for teaching mathematics but also someone who envisions children's mathematical learning in a similar manner as I do **AND** *someone who wants to take on the challenge. I have made a list of teachers in the school district with whom I have taught before but none seem appropriate for this particular project.*

That afternoon...

I scanned my INbox of emails and noticed that I had received something from my friend, Donna. I double clicked to open the message and found that she had written me a letter asking how my year of teaching had been.

Donna and I first met at the University while completing our Masters degrees in mathematics education. It had been two years since we graduated and we always talked about working together but the opportunity had not presented itself... until now. Not only were we seeking to teach in ways that focused on the growth of children's mathematical understanding, but upon reading the rest of her email, I discovered that she would also be teaching second and third grade children in the coming school year.

I double clicked on the REPLY button. Typing a response back to her as fast as I could, I told Donna what I had been up to in terms of my teaching and briefly explained my thinking behind the collaborative project. I then asked if she was interested in working together and clicked, SEND.

That summer and into the beginning of the school year...

Donna and I organized the math buddies program for our two classes. Using the theme of connecting our students with *their* mathematics, we focused on establishing learning spaces that not only emphasized the children's conceptual connections with mathematics but also, connections that would grow between the two classes and amongst the students as they worked together as one collective group.

Since the program centred on the children's mathematical interactions, we saw it to be a natural context for student-generated problem posing and problem solving projects.

Over the course of two years and despite the fact that Donna and I worked in different school districts, we co-created and taught mathematics classes simultaneously in our own classrooms and together as a group. We communicated to each other through e-mail, phone calls, and meeting each other in person. During these meetings, we reflected on lessons taught and took a closer, critical look at the kinds of teaching and learning that were arising. This gave us a chance to study the students' work and assess their mathematical understandings jointly, thereby, gaining insight from each other's perspective. It also allowed us to map out possible lessons and consider the impact that the teaching we proposed might have on our classes' learning. From here, we sketched out teaching-and-learning settings. Lessons were neither prescriptive nor planned too far in advance but evolved from the mathematics that surfaced in the classroom(s).

Keeping Our Classes Connected

Even when it was not possible for Donna and me to physically teach together, we took it as an opportunity to find creative ways to keep the children connected. One way that we did this was by teaching similar lessons within similar time frames. This provided occasions for sharing and examining the mathematics that was evolving in our classrooms. For example, as both of our classes worked on developing strategies for solving two and three digit computations, Donna and I presented work that children in the other class had produced. We invited our students to consider how the methods generated by their buddies were the same or different from theirs and whether (or not) and why the proposed strategies made mathematical sense. Not only did this enable the students to share what they were learning with the other class but it created spaces for them to critique each other's mathematics, to justify their thinking, and to offer new thoughts for their peers to consider.

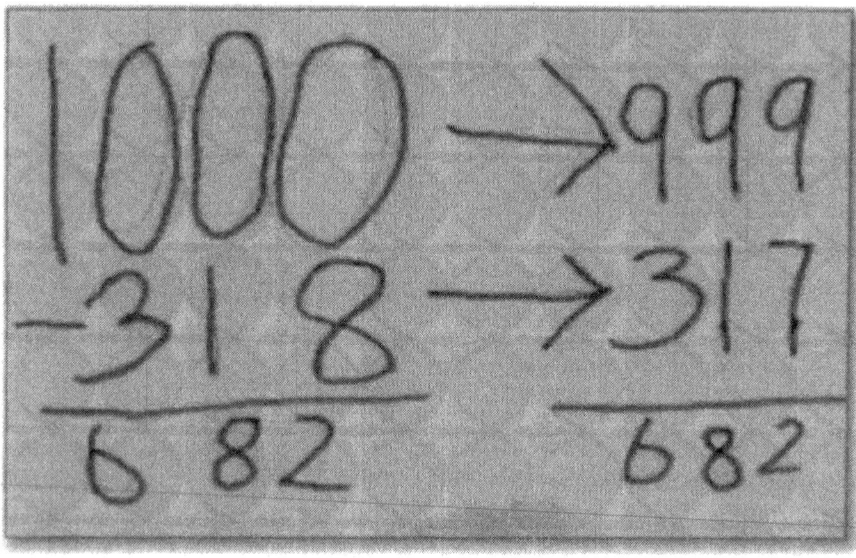

Robby's "take away one" subtraction strategy

Mathematical Gifts

The math buddies program was where the students' making and exchanging of mathematical gifts began. These gifts contained the conceptual ideas that the children happened to be exploring in class and by wrapping them up in imaginative contexts, the students' offerings turned into riddles, creative problems, and puzzles for their buddies to play with. Most often, the children worked with a partner during the posing and the solving of these projects.

Donna and I felt this to be important in order to maximize mathematical communication and collaboration.

The children's first gift to their buddies was a letter in which they introduced themselves. In class, the students had been investigating number concepts and decided that it would be fitting if they introduced themselves in mathematically imaginative terms using the concepts they were learning about, such as "I am in the grade that comes before four and two after Kindergarten" *and* "My favourite number is 12 less than 20 and 5 more than 3." *The children spent a good part of the next day measuring, collecting, and analyzing the numerical data about themselves such as their height, weight, age, shoe size, how many members in their family, favourite number, etc., and then developed mathematical descriptions for them. Donna's students were delighted to receive the letters. This in turn, opened up similar investigations of data collection and analysis for her students who responded with letters for my students.*

Reflect on the Reflection

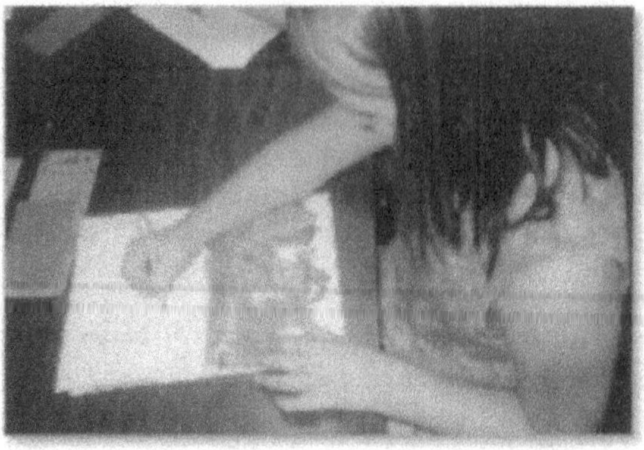

Danica drawing reflected images in the water

Robby: "Using a mirror helps me to check what I've drawn."

Reflect on the Reflection was a gift that came out of the classes' study of symmetry. Here, Donna's students painted pictures on one half of a piece of paper. They then gave them to my students who had to complete the other half of the image by drawing its reflection in the water.

> ***To solve this puzzle, the students needed to locate the horizontal line of symmetry and then with the aid of a mirror, 'flip' each of the images and paint them in on the other side of the line.***

Can You Build My Design?

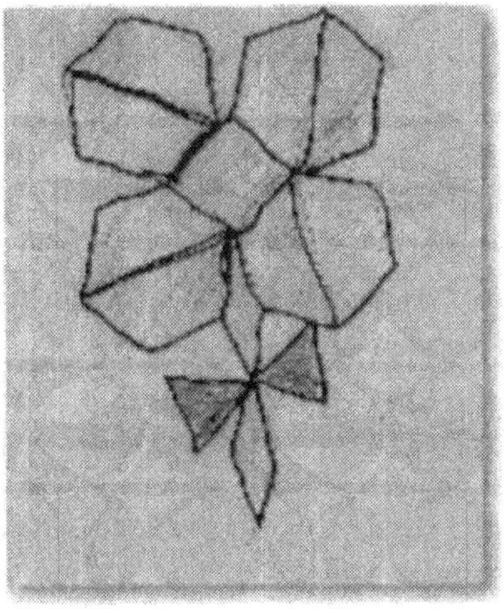

Flower pattern block design from Lara, Gregory, Mark, and Brian

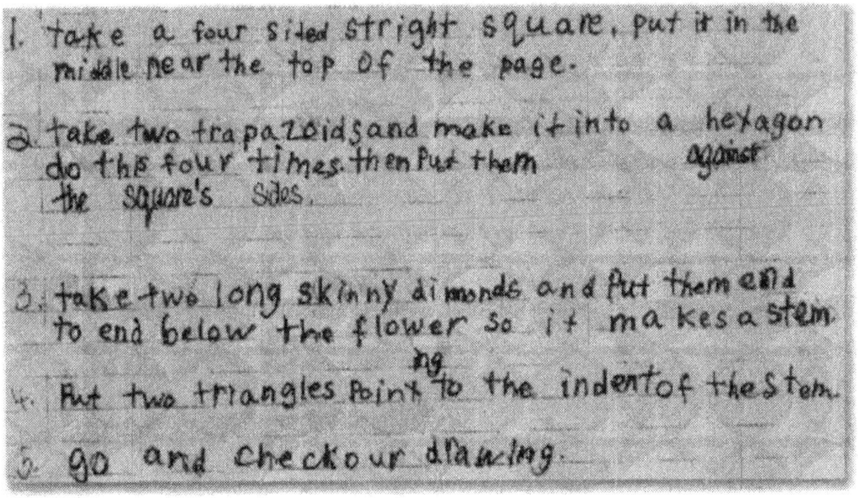

1. take a four sited stright square. put it in the middle near the top of the page.

2. take two trapazoids and make it into a hexagon do the four times. then put them against the square's sides.

3. take two long skinny dimonds and put them end to end below the flower so it makes a stem.

4. Put two triangles pointing to the indent of the stem.

5. go and check our drawing.

Lara, Gregory, Mark, and Brian's clues for their block design

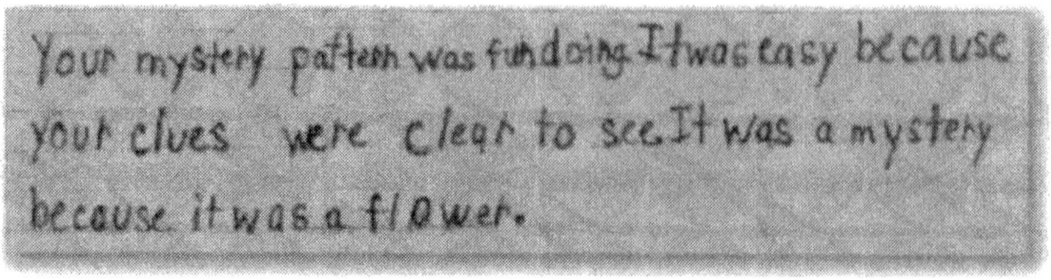

Your mystery pattern was fun doing. It was easy because your clues were clear to see. It was a mystery because it was a flower.

A letter from the group's math buddies

Can You Build My Design was a gift created by my students using pattern blocks to make a geometric design (e.g., a flower). Working in pairs or small groups, the children composed a written set of instructions telling their math buddies how to build the block design and included a picture of what it looked like so that they could compare it with what they had built.

> ***My students applied their understanding of geometry and spatial sense to describe, name, and interpret relative positions in space. For Donna's class who received the challenge, this puzzle required them to visualize, identify, and locate the described positions in space in order to build the design with pattern blocks. Once they had done so, the children compared what they had built with the picture of the intended design made by the creators of the problem and wrote back to their buddies identifying what part(s) of the instructions were clear to understand, what steps were difficult to figure out, and why.***

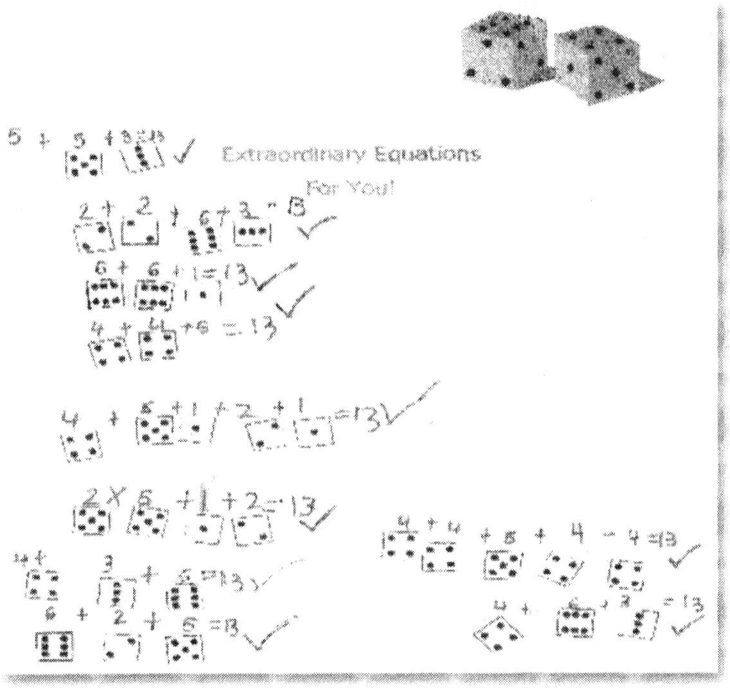

Carter's number stories about 13

(Let It Roll) Extraordinary Equations For You! was a game my students made for their math buddies. It consisted of a page full of dice showing values of 1-6 and a target number. Donna's students worked to use as many of the dice to create different number stories about the target number.

> ***The exchanging of gifts and sharing of work gave rise to mathematics and mathematical understandings that flowed back and forth, connecting our students and bringing our classrooms together. In this way, the teaching and learning that emerged in one classroom naturally became a part of what was taught and learned in the other classroom and so on. Very much like a conversation in progress, the mathematics and understandings that emerged through the continuous responses between our two classes provided an ecological space for the growth of our students and for us as teachers.***

Celebrating Together

As far as Donna and I were concerned, teaching mathematics was neither an exclusively independent activity nor did it need be restricted within the confines of one's classroom. In the math buddies program, one of our aims was to blur and at times, ignore or remove such boundaries we found to be limiting. In this way, we sought to relocate classroom mathematics by planning unil teaching lessons in tandem through sharing of students' work and in the children's exchanging of mathematical gifts. Another way in which we accomplished this was through what we called, celebration days.

Celebration days were when we brought our classes together and entire school days were devoted to engaging in mathematics as a collective group.

Getting acquainted with one another

The classes got acquainted with one another through a scavenger hunt designed for the children to learn about their math buddies' attitudes and thoughts about mathematics. The children asked students in the other class questions like, "What do you like about mathematics?," "What do you find challenging about mathematics?," ***and*** "Can you think of another way to describe the number, 10?" ***These questions were generated by the students, Donna, and me.***

Teaching our classes together gave Donna and me the opportunity to examine the children's mathematics *as* it emerged and discuss how to plan for subsequent activities. Because we shared similar views on mathematical understanding and considered it to be nonlinear, constantly growing, and fluid in nature, the mathematical explorations that we planned needed to be open and responsive to our students' thinking. This meant paying attention to the conversations and questions initiated by the children as they watched a film, read a story, or participated in an activity. Donna and I then integrated the ideas brought forth by the children into their mathematics for the rest of the day.

In the morning the children watched a film called Dance Squared and Donna had read a non-mathematical book titled, Selina and the Bear Paw Quilt. Donna and I were curious as to what would happen if we presented the film and the story one after the other and used them as springboards for the students' explorations that afternoon.

Sharing the story of *Selina and the Bear Paw Quilt*

Would the children work in pairs with their math buddies? Would our classes engage in some kind of collective, whole group mathematics? What would the mathematics entail? Geometric compositions of a square? Tessellations? Quilts?

The day before, Donna and I collected materials that we thought the children might find useful such as: wallpaper, rulers, scissors, pattern blocks, plastic 2-D shapes of the ones that appeared in the film and the story, chart paper, large pieces of felt material, glue, scrap paper, pencils, a class set of chalkboards, and coloured paper.

The questions and discussion that arose after watching the film and listening to the story created the context that shaped the afternoon's investigation. Over lunch, Donna and I assessed what we thought to be the main themes coming from the children's reactions to the film and the story. Their comments focused on the possible shapes that could be used in composing a square and the shapes needed to make the designs and pictures inside the quilt squares. For instance, "the shapes you would need to make the bear's claw." *Others spoke about the transformations of two-dimensional figures that would be necessary —* "how you'd need to turn the shape" *in order to make a particular geometric design or picture for a quilt square. And given that the entire discussion was directed toward that of quilts and quilt squares, this became the setting for a group project.*

That afternoon, the children decided that each of them would work with their math buddy from the other class to plan and produce one square that would eventually form a quilt made by the entire group. Donna and I explained the two themes that we had drawn from the group's discussion and asked the students how these might be worked into their quilt squares.

"The partners can decide what they want to do... they can do a design or a picture on the quilt square," said one student.

"But they have to do that... they can't just put shapes all over the square... that don't make sense," added another student.

"And, we can use the pattern blocks to figure out how to turn them... as tracers too," explained a third student.

The children dispersed from the middle of the gymnasium floor to find their math buddy and to gather the materials that they needed.

Donna and I made our way around to all of the students either individually or together. Watching, listening, and asking the children to explain what their work entailed, revealed three geometric and spatial strategies that the children utilized to plan and make their quilt squares.

Some students used paper and a pencil to sketch their picture and then used the pattern blocks by moving them about. By rotating, flipping, or using them in combination with other blocks, they were able to cover the regions of their image with appropriate shapes. Other children chose not to draw their design but used only the pattern blocks as manipulatives with which to plan out and then trace them to make their quilt square.

Danny and Jerry's pattern block design

Still, other students relied on their mental image or physical use of a single pattern block to perform transformations.

Clare explains, "If we use the triangle and keep turning it this way again and again until it gets back to the top, it'll make the petals of the flower.

Clare, Julia, and Sean's quilt square design

The students' finished quilt

By the end of the day, the children had created a quilt that integrated the designs and patterns they experienced through the film, the storybook, and their working with manipulative materials.

Number gymnastics

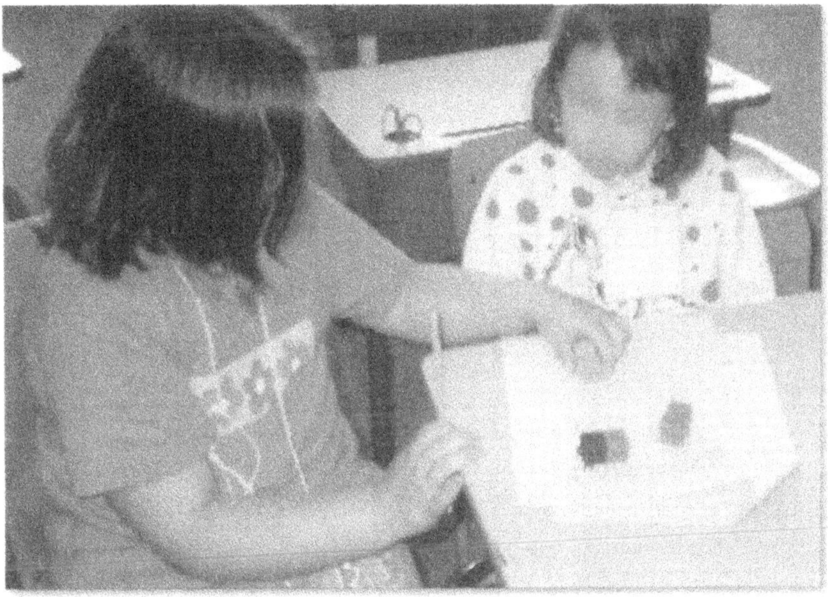

Showing Melissa why 5 is an odd not even number

During one of our other celebration days, the children helped to create and participate in *Number Gymnastics.* The students selected one of their favourite numbers and used it to work through a series of tasks that challenged the students' conceptual thinking and provided Donna and me with a chance to observe their flexibility of number sense of concepts we had taught during the year.

> ***The mathematical prompts came from a journal that I had kept during the course of the school year. In it, contained a collection of curious questions and observations made by Donna, the children, and myself.***

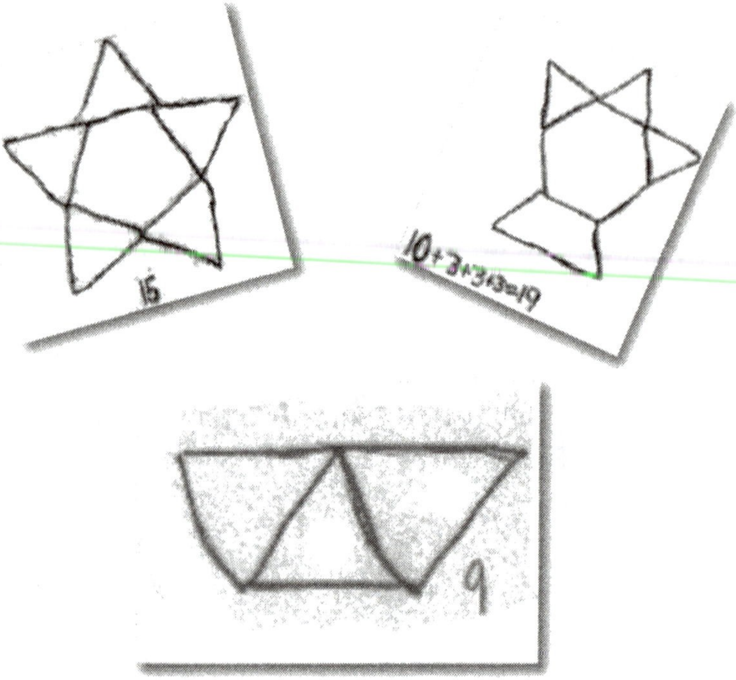

"Can you create and trace a pattern block design that has the same number of total sides as your number?"

"Look at your design. What fractions can you find?"

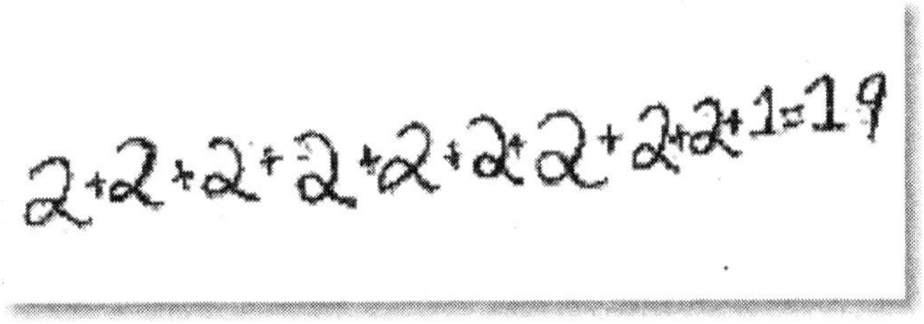

"Is your number odd or even? Can you show how you know this?"

Comparing thirteen and nine

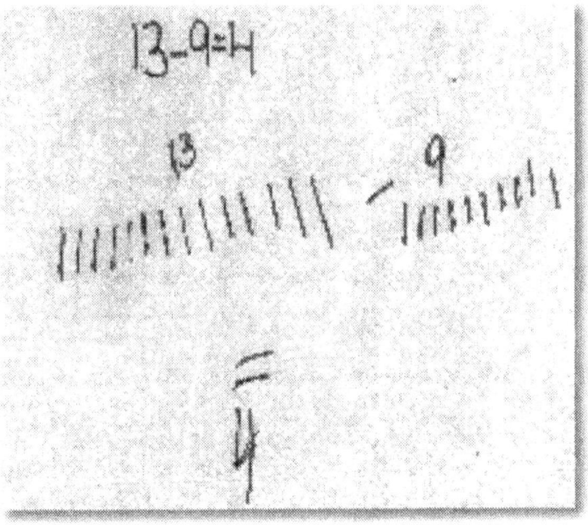

Comparing twelve and nineteen

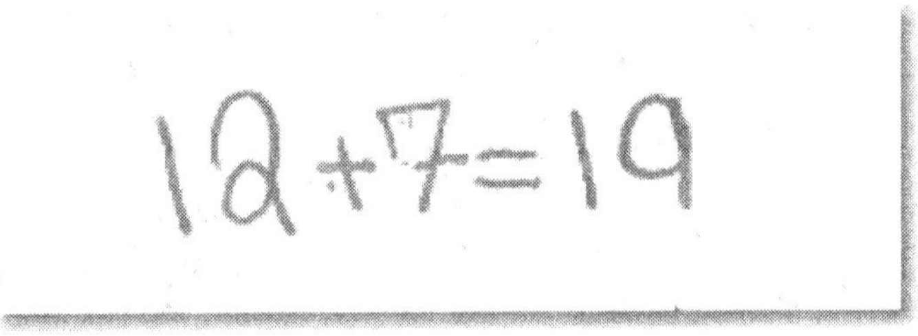

"Compare your number with__?? (you choose!) How much larger or smaller is it? Make a diagram or write with symbols to show how you know this."

Identifying equal "parts" of 15

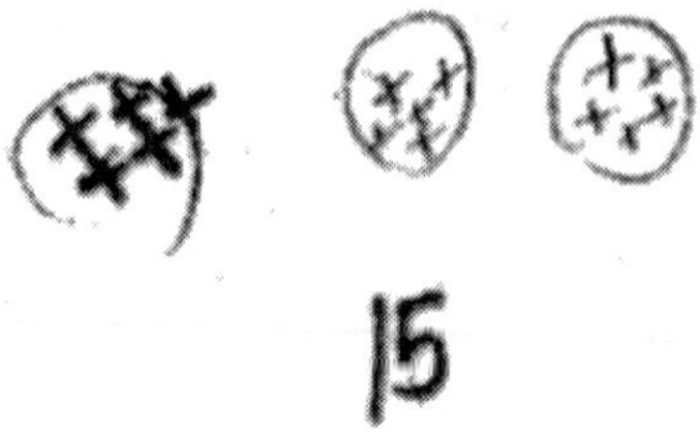

"Show how you can 'split' or 'divide' your number into smaller parts without remainders or leftovers."

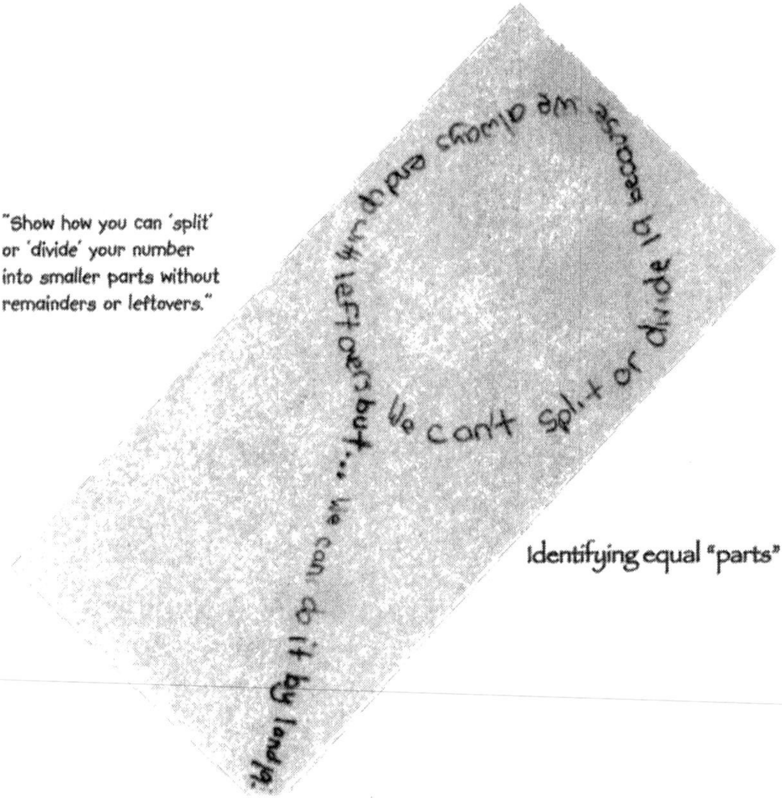

Identifying equal "parts" of 19

Beyond Spaces of Interaction

Mathematics is not something we have to look up to. It is right in front of us, at our fingertips, caught in the whorl of patterns of skin, in the symmetries of the hands, and the rhythms of blood and breath.

~ David Jardine[1]

Connecting Me, Connecting Us

Classroom mathematics beyond spaces of interaction? What do you mean?

I mean, attending and responding to children's mathematical learning by coming at it from a different direction. Instead of focusing on the emergence and development of ideas in relation to the children's interactions, it's about making spaces for them to see how their mathematical ideas and understandings inform who they are and the contexts of which they are a part.

So it's more about engaging children in thinking about how the mathematics they already know connects to them, to the world, and vice versa rather than their bringing forth of mathematics?

Exactly.

Okay. So why Jardine's thought?

It's another reminder that mathematics isn't something that's out there. It's right here ... quite literally, in you and me... our bodies as examples of mathematics... can't get more personal than that can you?!

Yes. Literally what it means to be mathematical.

So it was Jardine's idea that prompted me to see the need to occasion children's learning in a different way-- one that literally connected them to their mathematics.

"Are you a square or an oblong?" I asked the children.

"What?!" was the response I received from my students as they looked at me intrigued but also perplexed by the question I had posed-- what did geometry or more specifically, a square or an oblong have to do with them?!

Showing them a copy of the diagram:

I explained to the children that Leonardo da Vinci — whom we were learning about at the time — had described a perfectly proportioned person to be shaped like a square; that is, the distance of one's arm span is to be the same or close to one's head to toe measurement. After we discussed the 'squareness' of the body in the diagram, I turned the conversation to what it might mean if a person was not square but oblong. Giving the students squares and oblongs to physically manipulate, the class considered the criteria that defined a person as square and compared it to what they knew about oblongs.

The class made the argument that an oblong is different from a square in that when you turn a square so that it "sits" on a flat surface, it "*always looks the same*":

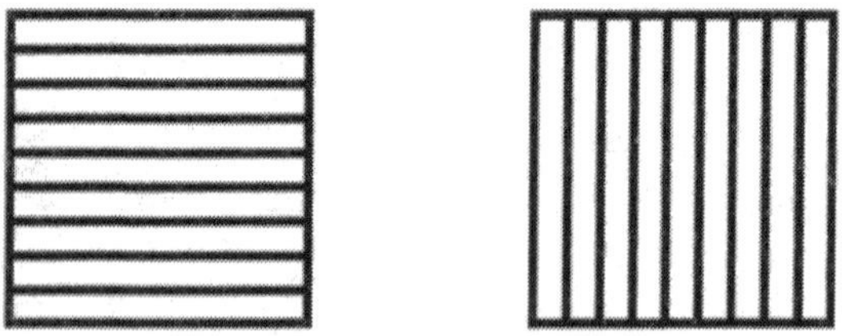

But when you do the same to an oblong, "it doesn't always look the same because oblongs have two longer sides and two shorter sides":

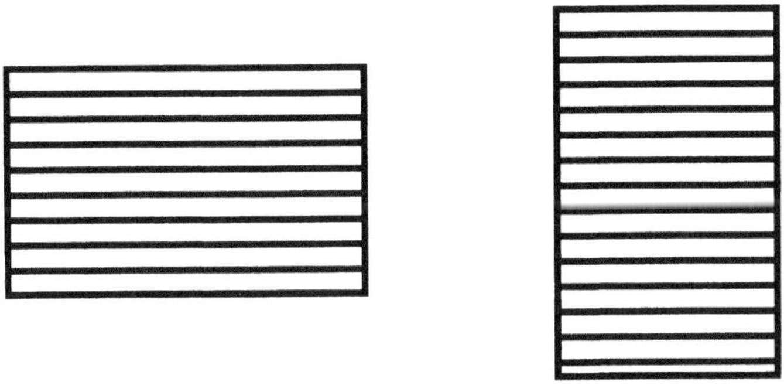

From this, the children made the conjecture that if a person was oblong, he or she might be a "wide", "side to side", or "left to right" oblong. This meant that the person would have to be "wider than they are tall." In addition to this, the students explained it was possible that another person could be a "tall" or "up and down" oblong. In this case, the person would have to be "taller than they are wide."

So, here.... squares and oblongs. These are the concepts the children were to think with.

Yes.

And from their verbal and physical descriptions, the children have conceptual meaning for some of what makes a square a square and an oblong an oblong.

And they've taken their spatial understandings-- dimensional properties about squares and oblongs and connected these to dimensions of the body in order to make distinctions between different body shapes.

The class then worked with a partner and carried out the study. They used a tape measure to find out their head-to-toe and arm-to-arm measurements, rounded the values to the nearest ten centimeters, and then identified what shape they were.

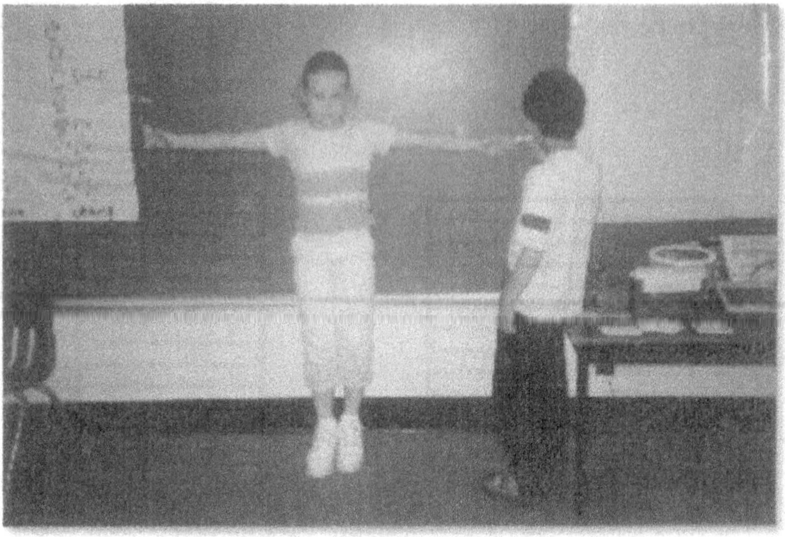

Robby and Madelaine: Are you a square or an oblong?!

Some found out that they were squares while other children discovered that they were tall oblongs. Of a total of 21 students, 9 were square and 12 were tall oblong. To the children's surprise, there was not one wide oblong in the entire class.

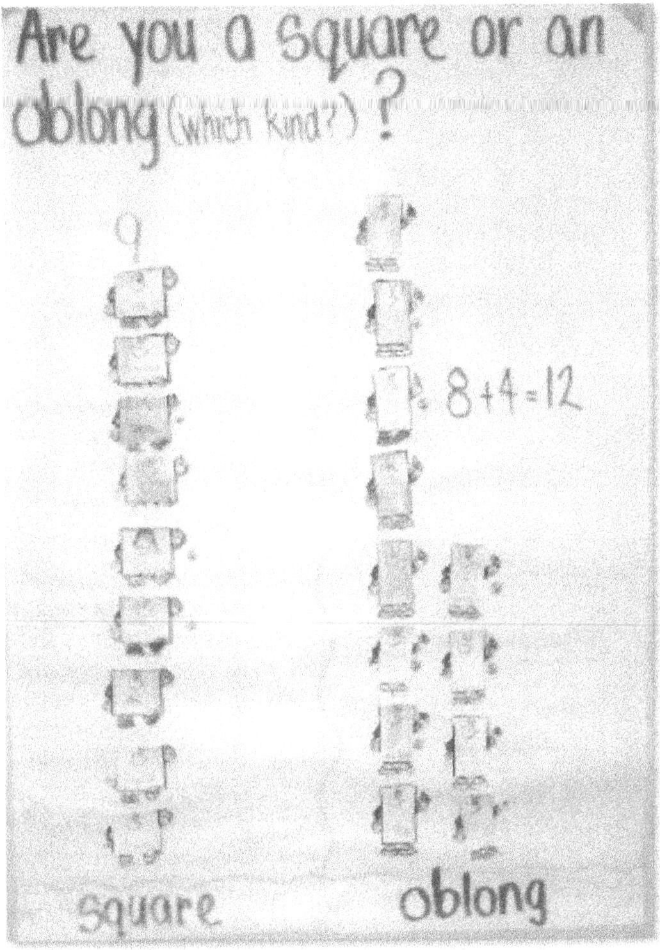

The class graph

Most people are tall oblongs.
~ Amie

There are 3 moer oblongs. then squares.
~ Madelaine

I was exactly 102 cm (arm span).... about 100 cm.
I was exactly 124 cm (head to toe).... about 120 cm.
I am [a tall] oblong.

~ Annie

I was exactly 118 cm (arm span).... about 120 cm.
I was exactly 129 cm (arm span).... about 130 cm.
I am a[n] up and down ob[l]ong.

~ Madelaine

I was exactly 120 cm (arm span).
I was exactly 120 cm (head to toe).
[I am a] square.

~ Sharon

My partner was Holly and she was a [tall] oblong because she is different both way[s].

~ Charlotte

Okay. So... yes, they were collecting the data. Yes, they were measuring and rounding numbers to the nearest ten. Yes, the children were analyzing the graph. But that's not what it was about was it? I mean, that's some of what they were learning, but the mathematical space that you opened was for the children to think with the idea of a square and an oblong in a context that hadn't occurred to them. In order for them to distinguish what 2-D shape they were, the children had to first spatially coordinate the two dimensions of length and width, and then compare these against the properties of the 2-D shapes.

Yes, that's right.

The following year we continued the investigation. The eight second-grade students who were now in third grade were curious to find out if their body shape had changed or remained the same from the previous year. The entire class wanted to know if there would be more or fewer oblongs than squares and how this year's results would compare to that of last year's data. Most of all, the children still wondered if it was even possible for someone to be a wide oblong.

So this space that you previously opened has now been re-opened and extended by the children themselves.

Yes, and now it included different contexts in which they could further examine the idea of square and oblong bodies. Their extension of the learning space gave rise to the class' posing of new questions.

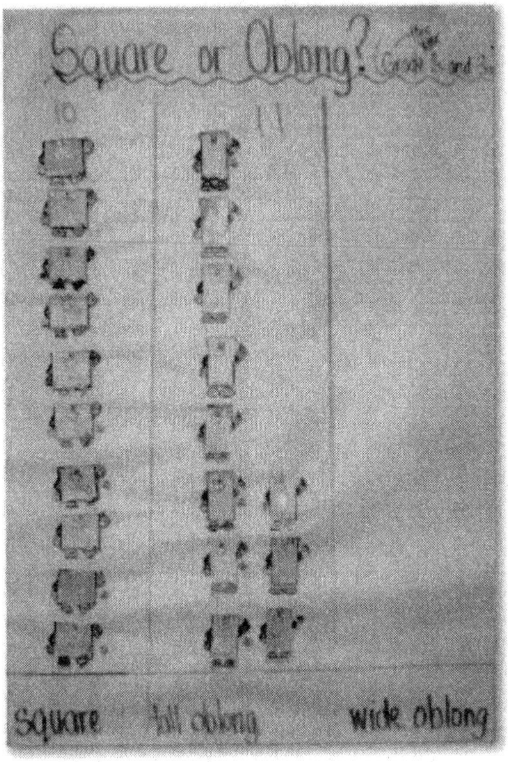

The class graph

Like last year, there were a total of 21 students. After the class collected their new data and analyzed it, they found that 10 of the 21 children were squares, 11 were oblongs, and still, there was not one person who could be considered to be a wide oblong.

Sitting in front of the two graphs, the children began comparing them.

Ellie and Michelle said, "Last year there were nine squares and this year there are ten squares. So there's one more square."

Mary-Jane identified that "with both of the graphs, there are always more tall oblongs."

Danica and Shelby made the claim that "most people didn't grow from last year. They just stayed the same."

This comment caused a great deal of talk amongst the class as they questioned whether it made sense. According to the students, each and every one of them knew for sure that she or he had grown in the past year.

When the class looked at the graphs again, I asked, "Can we tell how much people grew this year?"

Brian shook his head "no" but could not explain why. Shane argued that one would need to "compare the measurements from last year [with this year] and see how much it's changed."

The class came to the conclusion that even though the graphs provided information about the general shapes of students, the graphs did not contain the necessary details for making statements about people's specific measurements.

Picking up on Danica and Shelby's idea-- that most of the returning students had stayed the same shape, I turned the class' attention to another interpretation of the graph. Of the returning students and according to the first graph, two of students had been squares and six had been tall oblongs. This year there were four students who were squares and four who were tall oblongs. I asked the class if the statement that only two of the returning students had changed shape from last year would be an accurate one. Many students were quick to agree because of the difference of two between four and two and, six and four. Others felt that they should be more skeptical and indicated "maybe not" but could not explain why. When I asked the class what they would need to know in order to determine if students remained or changed shape from one year to the next, they said that you would have to find and compare the results of each student. So, each of the returning students located themselves on last year's graph and compared the results with the current graph.

So why did you do this?

Because, what Danica and Shelby said had caused such a reaction from the students, it needed to be brought around again so the class could consider the claim more critically... so that they could take the two sets of data and see what meaning they could make from them.

So, not just a superficial reading of the graphs. I mean, not just looking for the obvious facts like how many of this or that but given the fact that there are so many of this or that, what are the relationships that exist in the data, right?

Yes. Shelby and Danica's statement that most people hadn't grown from last year occasioned the need for the class to go back into the data, see what relevant information they could find, and then stand back and assess the claim by sorting what they knew and what connections existed between the two graphs. In other words, what was specific and lost in the graph such as their names and measurements and what was retained and could be used for comparison (i.e., their shapes).

Michelle saw that she was, "... the same thing as last year (i.e., tall oblong)." and speculated that she had probably grown in both her length and her width but was still taller than she was wide.

Madelaine giggled, "I was a tall oblong and I'm a tall oblong this year."

"I was an oblong last year and now I'm a square this year!" exclaimed Annie.

"I turned from a tall oblong to a square" said Robby.

Shane found that, "Last year I was an oblong and this year I'm a square".

This year Danielle measured as a tall oblong and attributed the change in her shape to the fact that, "I've grown more than my arms have (stretches out her arms). My legs have grown more because last year I was a square."

Holly said, "last year I was an [tall] oblong and I guess I was a little bit shorter and my arms weren't as long, and this year I'm a square."

As for Mark, last year he had been a square but "I'm a tall oblong this year. I've probably grown more in my head to toe measurement this year."

So by having them go back and compare the data from both years on the actual students themselves as opposed to just looking at the graphs' total number of squares and oblongs...

— They found out that in fact, only two of the returning students had remained the same shape: that is, tall oblongs while the other six students had changed. The exact opposite of Danica and Shelby's initial conjecture.

Exactly.

Still in search of people that fit the characteristics of a wide oblong, the children decided that such a body shape might not exist in the second and third grade and that they should go home and measure one of their parents because, perhaps older people were!

They've extended this mathematical space again by raising conjectures based on the trends they've located in the data they've collected on themselves.

Madelaine predicted, "more parents would be square-shaped than anything else."

Mary-Jane had a hunch that "maybe we'll find a couple of... wide oblongs." And the rest of the class anticipated that there would be more tall oblongs than anything else. That weekend, the children collected the measurements of one of their parents and brought the data to class the following week. With it, the class produced a third graph.

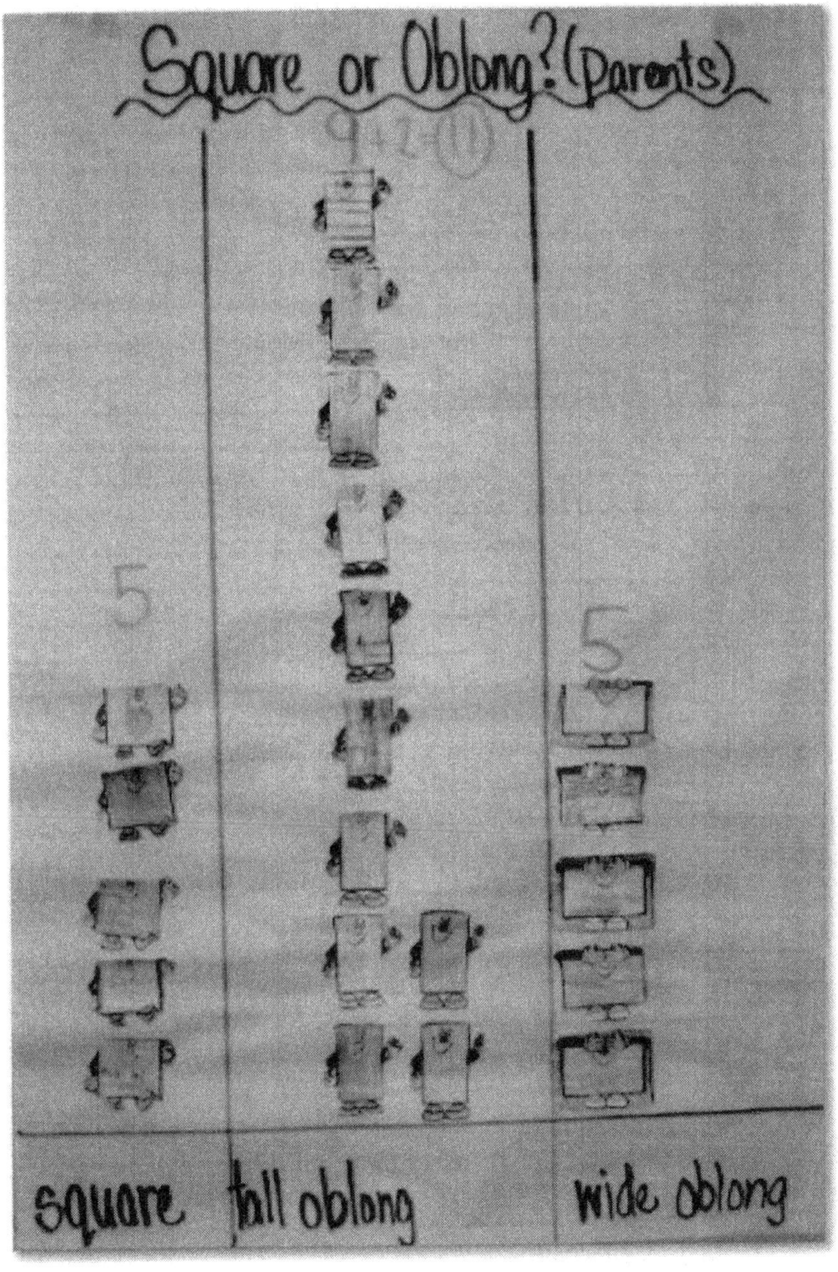

The graph, What's the shape of your parent?

[Parents that are] "square [or] wide oblong are the same." ~ William

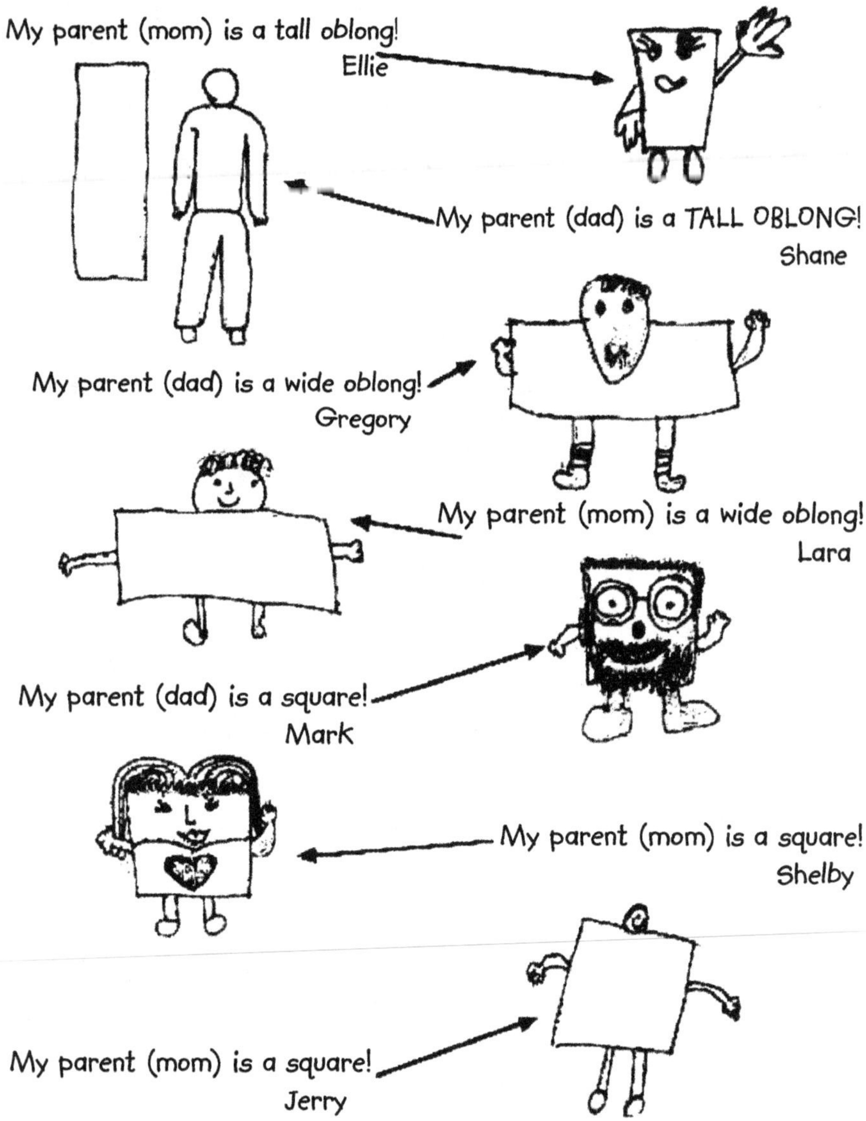
My parent (mom) is a tall oblong!
Ellie
My parent (dad) is a TALL OBLONG!
Shane
My parent (dad) is a wide oblong!
Gregory
My parent (mom) is a wide oblong!
Lara
My parent (dad) is a square!
Mark
My parent (mom) is a square!
Shelby
My parent (mom) is a square!
Jerry

We put this graph with the other two and examined it first by itself and then in relation to the other graphs. Madelaine's prediction that most parents would be squares did not prove to be true. However, the class' "guess" that there would be more tall oblongs than anything else was correct. The children were most excited to discover that Mary-Jane's hunch was a good one and that some parents were indeed, wide oblongs.

Mark then suggested that "you grow taller than your arm span does [as you get]… older."

"If that is true, why did we end up with [people who were] wide oblongs?" I asked.

Mark shrugged his shoulders and laughing, exclaimed, "maybe they shrunk!"

I asked the class what they thought might happen if we continued to measure and graph people's shapes. "Do you think we might… keep finding the pattern that Mark pointed out? That there are always tall oblongs than anything else?"

The students began talking to each other about what group of people they should sample next. Robby thought that the Kindergarten class would be a good group to survey to see if the pattern of tall oblongs would persist "because we've done older people-- our parents, and we've done medium people-- us, so we should do a smaller [i.e., younger] group."

So from the three sets of data that the class has collected, they've identified a pattern of tall oblongs that's consistent with all three graphs.

And now they're moving to a different age group to confirm it.

In addition to engaging the children to think critically and locate relationships in the data, another aspect of integration has emerged in terms of their mathematics. They are now collecting new data and interpreting it in light of the other sets in order to validate the patterns they've found, right?

Yes.

The class agreed with Robby's idea and Keefer predicted that the pattern would continue. Jerry figured that there would be more squares amongst the Kindergartens than any of the other groups surveyed.

Mark pointed out that one should not compare numbers without considering the total number of people in each group — "it depends how many people are in there [i.e., in the group]."

I acknowledged Mark's comment as being an important one for the class to consider. With all of the groups roughly the same in size (i.e., 21, 21, 19, and 20) the children and I decided that in this case it would not be a major concern.

All of the students predicted that we would definitely find wide oblongs in the Kindergarten group. When I asked them why they thought this would be so, Gregory raised his hand and explained his 'theory'.

"They're height is really small, the Kindergartens, they're pretty small-- up and down" he said nodding his head, "and they're arm span is probably more." The entire class agreed with Gregory's reasoning and on that note, we left the room to go and measure them.

What shape ARE kindergarten children?

Wondering...

and finding out!

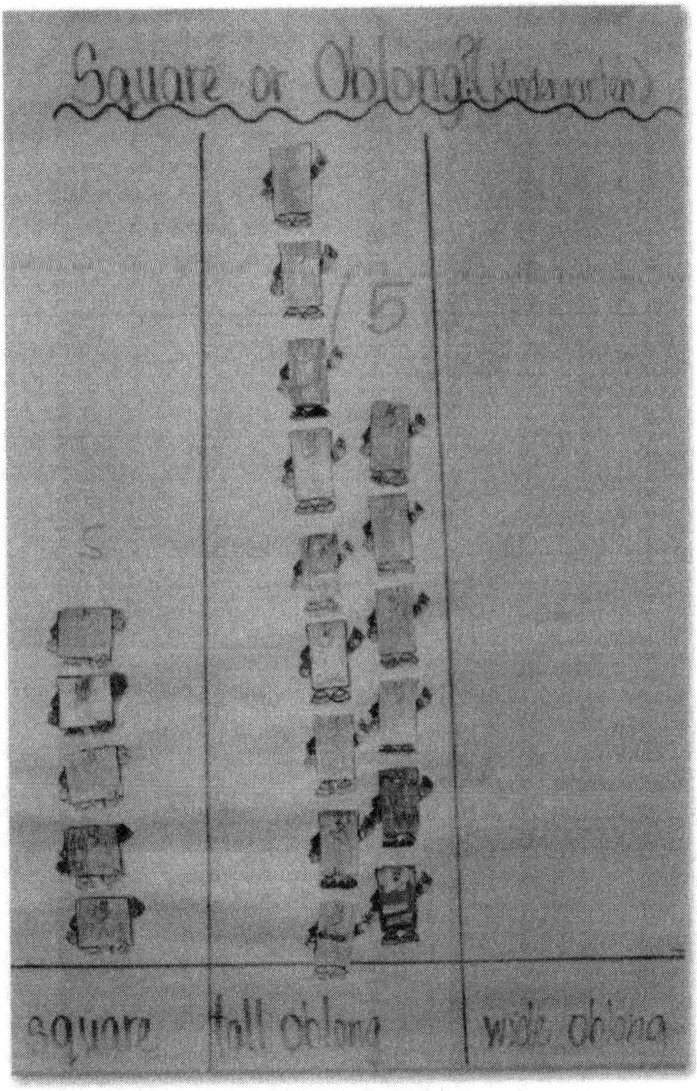

The graph of the Kindergarten class

When we returned to our classroom and hung the new graph with the others, Jerry looked at all four graphs on the chalkboard and said, "One was right and one was wrong." Here, Jerry's comment referred to the class' predictions.

"Which one was right?" I asked.

"The tall oblongs" he said. "We thought that they might have more squares but they actually have the same as the parents."

I asked the class, "What was something that was surprising about the Kindergarten's graph?"

Keefer raised his hand and said "We thought that there were less tall oblongs and we thought that there's more wide oblongs."

"And what did we find out?" I asked, prompting Keefer to continue.

"That there's actually more tall oblongs than wide oblongs."

Bringing focus to all four of their graphs and the *Are you a square or oblong?!* study as a whole, I asked the class that given the results of their work, if we cannot assume that Kindergartens will be shorter than their arm spans, what kinds of assumptions could we make based on the information they had collected and found in this exploration. In other words, did the mathematics connect us? Or, not?!

Here, making my reason for this mathematical investigation clear to the children...

...that the study was for them to develop complexity in their analytic thinking -- about how they can interpret data and locate meaningful patterns by working across sets of data.

"People really do come in all shapes and sizes!" said Madelaine.

A VERY general pattern!

"No kids are wide oblongs," said Gregory.

"Every graph, there are more tall oblongs than anything else," explained Jerry.

And Brian pointed out that according to the graphs, the group that had the most square people were children of the second and third grade. One final mathematical connection that emerged for the class was a pattern about how people's bodies grew.

Robby explained that in Kindergarten, "you're more tall than you are wide."

Sarah added, "your arm span grows bigger and it evens out with your height," as you get older.

The class then concluded that the most variety of body shapes happens when you become an adult. You either remain square,

"Or you grow taller [i.e., a tall oblong]," said Mark.

"Or your arm span goes longer than how tall you are," said Robby.

And here, examples of specific relationships or trends that the class located across the collected data.

I can see now how this learning space that you opened by having the children take already known geometric concepts and apply them to a new context enabled your students to not only to think in mathematical ways but to do so in a generative manner. Through the thinking that the children developed, they experienced how their mathematical ideas and understandings informed who they are and connected them with the contexts of which they are a part. From the questions and conjectures they raised about certain patterns that might exist within and across groups of people naturally created the need for your students to plan and generate different sets of data that lead to their activities in critically analyzing the data... that occasioned new insights... and in a recursive way, gave rise to new investigations. Your students connected a real-life phenomenon--how people grow, to explain their mathematical findings. In essence, bringing the self, the world, and mathematics together. Mathematics wasn't just a set of facts but directly related to them.

Connecting Us to It

In my work to embed an ecological sense of place for mathematics by opening spaces for children to interact and connect with mathematics, I cannot neglect the spaces in which mathematics connects us to it. By "it", I mean the more-than human world.

Another of what you'd call, a free idea?[1]

Yes. Thinking with Gregory Bateson's notion of patterns which connect and relating it to what Abram speaks of as the more-than-human world.

You're speaking of the natural world?

Yes, but not in an objectified or a disconnected manner... an environment that we are connected to and with which we interact.

Right. And so, how do Bateson's and Abram's ideas come into play?

In making the need for spaces in which students interact mathematically with the natural world explicit and not for the sake of discovering universal truths or facts.

You mean, thinking mathematically and forming relationships through these interactions.

Yes, exactly.

This connects with what you've quoted earlier-- from van Manen — it's about enabling children to know the mathematics in a way that it speaks to them. Here, through contexts which are other than human.

And how they might develop mathematical ways to speak to it... to respond in such settings.

Like what?

T*aking a Closer Look at Snowflakes* was a theme I developed to connect the children with the natural world in mathematical ways. Because my teaching was grounded in the idea that learning can arise from reconsidering that which we take for granted as known terrain, the idea of my students reexamining the familiar snowflake seemed fitting.

With a few snowfalls, the children's dark clothes, magnifying glasses and incorporating films, photographs, and books about snowflakes, the "small white stuff" became fascinating and complex mathematical pattern makers.

What kinds of patterns?

Well, the class was surprised that there could be so many different geometric shapes in snowflakes — "diamonds", circles, pentagons, trapezoids, hexagons, squares, oblongs, triangles, "skinny" ovals, flower and star shapes... Here's a diagram made by Isa:

Isa's close-up study of a snowflake

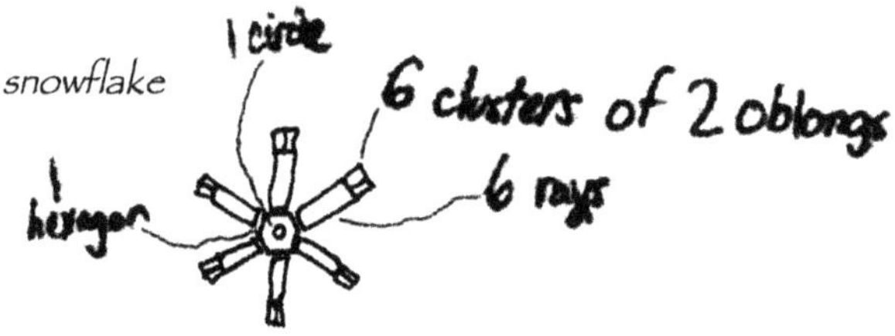

Not only has she identified the geometric shapes within the snowflake but also, the particular organization of them.

This was something that intrigued the class-- the actual number patterns in the arrangement of the snowflakes. As the children located new ones, they called them out and I recorded them on a large piece of chart paper.

6 sticks	1 hexagon	12 oblongs
6 diamonds	1 star	12 petals
6 circles	1 flower	18 squares
6 pentagons	2 hexagons	18 points
6 trapezoids	120 icicles	12 triangles

6 groups of 2 triangles (6x2) (2+2+2+2+2+2)
6 groups of 4 triangles (6x4) (4+4+4+4+4+4)
6 groups of 3 rays (6x3) (3+3+3+3+3+3)
6 groups of 3 points (6x3) (3+3+3+3+3+3)
6 groups of 2 squares (6x2) (2+2+2+2+2+2)
6 groups of 2 oblongs (6x2) (2+2+2+2+2+2)
6 groups of 3 circles (6x3) (3+3+3+3+3+3)

Snowflakes as geometric and numerical pattern makers

Once done, the class looked at the chart to see if they noticed any other patterns. Shouji made the observation that the number of shapes was always either one or an even number. Charlotte added to his statement by explaining that each of the snowflakes had structures that were "different but all groups of six." This she conjectured, was most likely because snowflakes begin with a six sided germ crystal.

Hence, the multiples of six that the students were thinking of in terms of repeated addition, 'groups of', and multiplication. Having now seen these shapes and patterns, in what direction did this study then move?

I challenged the children to create a mathematical response that would capture the dynamic qualities of their particular snowflake. Initially, the students began with verbal and symbolic descriptions that then turned into mathematical poems.

Poetry? In mathematics?

Yes. From the films we had watched, the children learned that a snowflake grows outwards from its centre— the germ crystal — that Charlotte spoke of. This is why the students always described the snowflakes from a perspective that began in the centre and moved outwards. In this manner, the children visualized the formation as an animate and artistic process... poetic in nature. It only made sense then, that their responses be poetic in form. And *reflected in each of them was the student's spatial and numerical thinking.*

I small circle in my center
I of my circles is around another circle
6 small triangles [on] the edge of my largest hexagon
I hexagon around my 6 triangles
I medium size hexagon between a big hexagon and a
  small hexagon
6 clusters of 2 squares in my big hexagon
I big hexagon around 2 small hexagons
6 triangles around a circle
6 rays in my middle
6 clusters of 2 forming a ring
I am telling you this while I am dancing in the air.

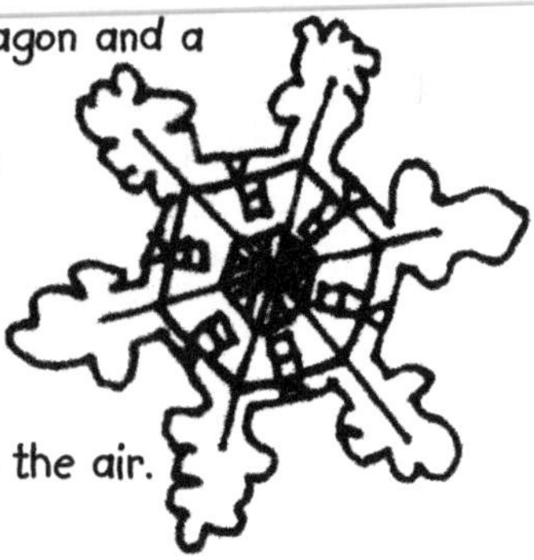

by Isa

I Flower with 6 petals,
I Star with 6 points,
I Hexagon,
12 Little triangles,
6 Big triangles,
Twirling in the garden.

by Sharon

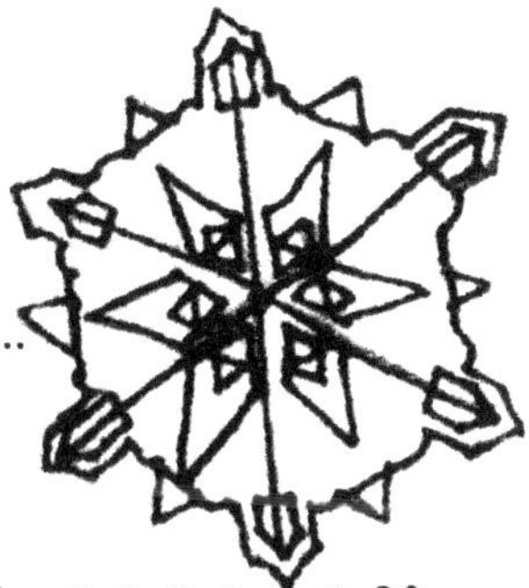

I circle in its centre
two rings of six diamonds around its centre
6 branches from the middle to the edge
6 triangles around its outside
12 bumps on the edge of its largest hexagon...
6 clusters of 2
bumps... 6x2=12
6 pentagons on the tip of its branches
6 clusters of 5 points along its edge... 6x5=30... 5+5+5+5+5+5=30
filling the air with other dancing snowflakes.

by Shouji

Through their poems, the children communicated the specific shapes, their locations, and the number relationships that they observed in their snowflakes.

Like in Isa's poem when she wrote, "6 clusters of 2 squares in my big hexagon."

Or when Shouji wrote, "6 clusters of 5 points along its edge..." and then echoed it two more times but in different ways through his expressions of 6x5=30 and 5+5+5+5+5+5=30. Each of their poems are responses in terms of how the snowflake as a mathematical form spoke to them but spoke in the poetic language of clusters, flowers, rays, and branches.

I yelled out: "You have to decide *now* which you are-- a GIANT, a WIZARD, or a DWARF!"....

"Where do the Mermaids stand?"

Where do the Mermaids stand?

A long pause. A *very* long pause. "Where do the Mermaids stand?" says I.

"Yes. You see, I am a Mermaid."

"There are no such things as Mermaids."

"Oh, yes, I am one!"

She did not relate to being a Giant, a Wizard, or a Dwarf.... She took it for granted that there was a place for Mermaids and that I would know just where.

Well, where DO the Mermaids stand? All the "Mermaids" – all those who are different, who do not fit the norm and who do not accept the available boxes and pigeonholes?

~ Robert Fulghum[1]

Spaces for Unpredictable Mathematics

Having experienced moments of being this little mermaid myself, I am trying in my praxis of teaching to be mindful not to pigeonhole anyone or anything. Given an ecological mind-space, all that I have explored and everything I have come to make sense of, it makes no sense that definitive categories or endpoints in children's mathematical explorations would be desirable or even realistic. If mathematics as well as the teaching and learning of it are to be fluid and responsive, then it is incoherent to assume that children's ways of being mathematical, and this includes their solutions, would be rigid and absolute.

In making space for children to adventure in their mathematical learning means that there is always to be an element of unpredictability and any endpoint remains as an open-closure.

> The teacher will often be in the position, unusual for mathematics teachers and uncomfortable for many, of *not knowing*; to work well without knowing all the answers requires experience, confidence, and self-awareness.
>
> – Hugh Burkhardt[2]

It also entails that

> ... teachers must perceive the implications of the students' different approaches, whether they may be fruitful and, if not, what might make them so.
>
> –Hugh Burkardt[3]

Is A Half Of A Half *Really* A Half?

The class had been busy working on making halves. Jennifer had given them a variety of everyday items such as collections of objects, dollars and coins, two-dimensional shapes, and containers of dry and liquid materials. The children worked in groups of three and four to find as many different ways of 'halving' the items and justifying why despite the different appearances of the halves, each of the items were in fact a half of the original whole.

It was just as the last group was about to finish sharing what they had found in their investigation of a half that Jennifer looked around at the rest of the class and noticed that Sammy was sitting very still in a hunkered down and slouched position, staring off into space– lips pursed and concentrating very hard... on something. Before she could ask him what he was thinking about, Sammy sat up, leaned his body forward, and pointed his right index finger towards the ceiling.

Looking first directly at Jennifer and then to his classmates with wide eyes and a sense of urgency in his voice, Sammy sputtered out, "I wonder... is a half of a half a half?!"

For a brief moment the class was completely silent. No one said or did anything. And then, just as quickly as the children had become quiet, all of a sudden, they were abuzz-- turning and looking at one another with furrowed brows and asking, "is a half of a half a half?" Shrugging their shoulders, they looked to Jennifer for her response. Recognizing the mathematical playfulness of Sammy's question, she simply raised her hands, shrugged her shoulders, and tossed it back to them.

"*Is* a half of a half a half?!" she said to the class. Jennifer was curious to see the ways in which the children would move inside this unpredictable space that Sammy had opened.

The class dispersed from the carpet and formed small working clusters around the room. Jennifer moved about, helping each group to gather needed materials so they could begin to explore the question. As she worked with the students, she observed that some groups were busy sketching out diagrams on pieces of paper while others were having a conversation and talking about what a half of a half was. Still other students took hold of actual objects– boxes, containers, sheets of paper, and geometric shapes and then proceeded to draw imaginary lines to make a half of a half. Emanating from all of this activity was a roar of, "yes it is!" and, "no it isn't!"

After some time, the class came together and each group presented what they felt was a convincing response to Sammy's question. Justifications for why a half of a half was a half had to do with the fact that when something is divided into two equal pieces and each one of the pieces is a half of the original half piece. There were other groups however, whose explanations pointed to the fact that when you divide and take a half of a half, what you have taken or what is leftover is one quarter– *not* a half.

Consequently, the class remained divided-- in half. Taking a hold of Sammy's question and through several acts of mathematical prodding, tugging, and pulling apart, the children arrived at two points of viewing and seemed adamant that a half of a half had to be one or the other. Either it was a half or it was not.

Sammy sat quietly on the carpet for the entire discussion, looking and listening to what his classmates presented. When the conversation came to a grinding halt, Sammy sat up as he had before but instead of opening a mathematical space, he jumped into the thick of it and brought the two confounding interpretations together.

"It is and it isn't," Sammy said. "If you're only looking at the half," he said, "then it is a half. But if you are looking at the whole then it isn't, it would be a quarter. It depends on how you are looking."

By pointing out that it *depends* on how the half of a half is imagined, Sammy effectively transformed it from being a *thing* into a mathematical *relationship* that is contingent on the context in which it is situated. The understanding which his shift in thinking enabled was that even though a person might take an either-or approach and statically define what a half of a half is (or is not), there is also the opportunity to consider in a systemic manner, what the half of a half is *part of* (i.e., the half or the whole); in other words, how it is related. By doing so, Sammy made it possible for the class to conceptualize how a half of a half could be both a half and a quarter.

One day, two years later, I found myself sitting on the pessimistic edge of skepticism. It was in this space of mind I began thinking that perhaps there are only wizards, giants, and dwarfs in the mathematics classroom. Maybe there are no such things as mermaids. Next thing I know, I am revisiting Sammy's question. It was as a little test, if you will, to see if such spaces really are as curious and unpredictable as I think them to be. If indeed they are, then even when reopened, one cannot be certain what will happen *this* time.

In Search Of Mermaids

I gathered up my students and took them to Sammy's space.

"That's weird!" said Robby as he and the class met the question, *Is a half of a half a half?*

> ***Like the students two years before them, the idea of a half — something so everyday and familiar, suddenly became strange and unfamiliar.***

The children did not need any coaxing from me to enter this mathematical space. They jumped right in. Once inside, the class proceeded to take a good look around.

"Is a quarter of a quarter a quarter?" asked Danny.

"Is a third of a third a third? Is a tenth of a tenth a tenth?" said Clare.

> ***The children were enchanted by the idea of a fraction of a fraction and the linguistic rhythm of it appealed to them too. From here, the class went on to generate several queries.***

Julie wanted to know if a whole of a whole could still be a whole while Ethan wondered about a sixth of a sixth, and Shouji rounded out the list of questions by asking whether or not a fifth of a fifth really was a fifth.

I recorded the children's questions onto a large piece of chart paper and after some discussion as to how we should continue, the class decided to break into smaller groups of three or four and each group chose a different question to investigate. The children also decided that it would be best if they all used a circle shaped pizza to be the common 'whole' and for each group to make a poster so that the results of their work could be displayed.

When all of the groups had finished their posters, the class sorted them in order from the greatest fraction-- one whole pizza, to the least that was the tenth of a tenth of a pizza. We then taped the posters up across the length of the chalkboard so that everyone could see the entire class' work.

Julie's group arrived at the conclusion that a whole of a whole could not be anything but a whole. Shane's group took the position that a half of a half was not a half but rather, a quarter, because the resulting slice of pizza was one of a total of four pieces.

Wanting to engage the students' thinking further, I asked the children if they agreed with Shane's group's presentation that a half of a half is always a quarter. Robby nodded his head in agreement that a half of a half could be considered to be one quarter as Shane's group had demonstrated but then pointed to the group's poster and also explained that, "from a half, it is a half."

> ***What initially appeared "weird" to Robby was now something curious and playful. Like Sammy, Robby too had arrived at a place of knowing where a half of a half could possess two identities. But unlike the class two years ago, this class was dealing with several fractions of fractions and with one that happened to be one whole. And so, Robby's thinking did not become a residual understanding[4] as Sammy's had, but a source[5] that set the class off on a comparison of all seven posters.***

The students agreed with Robby that the half of a half could be viewed in relation to the whole or the half, but they also argued that a whole of a whole could not be anything but a whole because the one pizza was its only reference.

The class moved on to Danny's group's poster. Danny explained that a third of a third could be a third because, "if this is one third (pointing to one of the three pieces of pizza)... and you split it [i.e., into three pieces again] it's like splitting a package into a third."

Isa stared at the resulting third of a third, shook her head, and then, said nothing. Asked to come up to the poster and show the class what she was thinking, Isa offered a second interpretation. With her hand, she covered up the thirds of the pizza one at a time and pointed out, "there would be three, six, nine pieces. It's one-ninth."

Ethan said that, "A quarter of a quarter is a quarter because it's one fourth of a quarter."

And Danny added, "four plus four equals eight and eight plus four... twelve, and twelve plus four is sixteen... a sixteenth."

Mac shared what his group had found out. "Yeah, I think it could be a fifth. You could call it a fifth and it's a fifth of a fifth. So it's just like pretending a fifth is like a whole of a pizza." He walked up to the poster and said, "If you look," pointing to the piece that was a fifth of a fifth, "and you... if you take all the pieces" meaning the entire pizza, "there'd be twenty-five little pieces, so it'd be one twenty-fifth."

Mark nodded his head and followed by explaining his group's poster in a similar manner; that a sixth of a sixth was a sixth "because there's six of them in the sixth."

The conversation was gaining momentum when Mac and Steven made their way to the front of the class. Mac pointed to the pizza and said, "There's eighteen [pieces] in one half!"

"And then eighteen on the other side," added Steven.

I joined in, "and eighteen plus eighteen is..."
"Thirty-six!" chimed Steven and Mac.
Smiling, Mac looked at the class and announced, "One thirty-sixth."

The class' working through the six of the seven posters in this way and recording the complementary fractions firmly established the two different ways of conceptualizing the resulting fraction: First, as a part of the original part, and secondly, as a part of the greatest whole.

It was when the class reached the last poster that something different and for me, unexpected happened. Robby and Danny raised their hands at the same time and both called out that if the entire 'tenth of a tenth' pizza was cut into slices that were all tenths of a tenth, there would be one hundred pieces in total.

Mark said, "There are ten groups of ten and so you have to do ten times ten."

Danny told the class that "if you put this altogether" – referring to the ten tenths of a tenth of the pizza, "that would make one tenth" of the whole pizza.
"So that would be another tenth..." he pointed out as he moved around the pizza in a clockwise manner, one tenth of the pizza at a time. "Until you reach to here."

By "here", he meant back to the place where you started.

"And you would find," Danny said, "that there would be ten of these", pointing to one group of tenths with his finger. "And then there'd also... they're tenths. So, you would find out that there'd be ten of them so you can think of them as ten times ten... they are tenths... so if you split them up like that and you have to have ten pieces to make a whole... one hundredth."

It can be assumed that Robby, Danny, and Mark, who, most clearly demonstrated this, applied their learned knowledge of whole number multiplication to this context of fractions. They thought with a 'groups of' notion to account for the total number of pieces (i.e., ten groups of ten pizza slices is one hundred slices altogether).

Just as none of the children had received any formal teaching on identifying fractions within fractions, they also had not worked on creating equivalent fractions, adding, or multiplying them. Yet, these were the understandings that emerged for Danny.

In his second response, Danny conceptually communicated that putting together (adding) ten one hundredths of a pizza results in a same sized piece (equivalency) as one that was one tenth of a whole pizza. He then demonstrated that in order to determine

the size of the individual pieces that result from sectioning off each tenth into tenths again, you multiply the tenths (10) by the number of groups of tenths (10) that now exist (10 x 10). This produces the total number of individual pieces in the entire pizza (100), and of which, each piece is one hundredth of the whole pizza (effectively, the reciprocal).

I was about to ask the children to go and get their lunches when Clare smiled and excitedly waved her hand back and forth.

"Yes, Clare" I said.

Beaming, it was obvious there was something important that she wanted to share with us. Clare came up to the front of the class, pointed to the fraction denominators on the posters, and beginning with one half and moving through to one tenth, announced:

"Two times two is four, three times three is nine, four times four is sixteen, five times fives is twenty-five, six times six is thirty-six, and ten times ten is one hundred" she chanted. "The numbers work" she concluded, matter of factly.

Upon first glance, one might suspect Clare had simply memorized her multiplication facts and recognized them here as a numerical pattern.

I encouraged her to continue. Clare began with the third of a third and then moved to the poster of a tenth of a tenth, She told the class that the reason why the numbers "worked" was because the piece, for example, the *tenth* of a tenth was becoming increasingly smaller.

Clare also pointed out that "if this one," she said referring to the pizza already divided into two halves, "is split in half, then there'd be one group." Clare showed this with her hands cupped around half of the pizza. Here, she demonstrated that "one group" meant one group of two pieces of half of the pizza, "and one group" she said, being the other half of the pizza that also had two slices.

Moving on to further prove her point that the pieces of pizza were becoming multiplicatively smaller, Clare went through the remaining posters in the same way she had done first and verbally highlighted the repetitive additive pattern that was connecting them all.

"There's three, three, and three, ... four, four, and four, and four, ... five, five, five, five, and five, ... six, six, six, six, six, and six, ... ten, ten, ten, ten, ten, ten, ten, ten, ten, and ten."

It was evident that Clare was not merely applying remembered facts but rather, thinking simultaneously with spatial and numerical ideas. The multiplicative pattern that she identified as "two times two is four, three times three is nine..." is related to

the operational and spatial action of making the pieces of pizza repetitively smaller as evidenced in "three, three, three" and "four, four, four, four..." and so on. Clare also connected repeated addition to multiplication; for example, "three, three, three" with "three times three is nine."

Looking back at the mathematics that defined this and, any of the other learning spaces, it could be argued that nothing extra-ordinary happened. In re-rooting the learning space, systemically ecological ways of thinking and doing mathematics had become the class' way of being mathematical.

This was indeed very different from what was previously taken for granted in my teaching. Clearly now was the assumption that the work to be done, the ways in which to accomplish it, and the ideas and understandings that were to be realized could not be predicted but only existed in the enactment of bringing mathematics into being.

In the classroom now, making sense of mathematics such as whether a half of a half is really a half had less to do with 'what is it?' and more to do with 'how' the mathematics — this creature — spoke to the children; how mathematics related to them and how they related to mathematics. Posing questions from a question, moving into, out of, and amongst spatial and numerical realms of thinking as well as arriving at different yet just as compelling places of knowing did not come so much as a surprise but more so, as an expectation.

So...

It's not true... that mermaids do not exist.

~ Robert Fulghum[6]

Off The Trail

SPACE WANTED

Looking to share a space with ecology. Interested in what ecologically coherent forms of teaching and learning mathematics could mean for the classroom. Can move in IMMEDIATELY.

HELP WANTED

"How can we break out of our conventional approaches and imagine more productive alternatives?"[1] Reply to mailbox: T1I9M9M7S

VACANCY

Seeking one teacher to teach grades 2/3. Separate room. Shared facilities.

(continued from page 36)

3 Faces of Ecology

According to M. C. Bateson[2] there are three "faces" or realms of ecology: empirical, enviromental, and systemic. The author defines empirical ecology as biological, meteorological, and geographical studies that focus on understanding how the planet is changing and how these changes affect the interrelationships of the world's natural systems. The environmental face of ecology is concerned with identifying the level of impact that our ways of living have on the earth's systems. This area of ecology also involves the development of solutions for environmental problems that minimize harmful stress on the earth. It is within the systemic realm of ecology where mathematics teaching and learning can be most radically explored. This is because systemic thinking focuses on seeking "the pattern which connects"[3] a system or systems together as interacting wholes.

A "system" could be a teacher or a student. It could also be a collective group such as a mathematics class, a school, and so on.

The connecting pattern or patterns that interrelate one system to another encompass the forms of knowledge, actions, and identities that result from the ongoing interactions within the systems as well as the ways in which these dynamics are sustained by the systems themselves.

Focusing on relational qualities makes it possible to view the world as an integrated whole; a dynamic and fluid unity in which all living and social-cultural systems are interconnected. The world is not conceptualized as a collection of separate entities but instead, as a highly complex unity in which all systems are interrelated and therefore, interdependent.

It seems sensible that when looking from an ecological perspective, mathematics teaching could be conceived as similar to that of students' mathematical learning.[4] Mathematics teaching then, as a fluid and complex process, implies that it exists always in relation to the ongoing interactions of the students, the mathematics, and the material and nonmaterial environment of the classroom.

In light of this, it is from a systemic perspective that the following question is posed:

> *In what ways can systemic manners of thinking about mathematics education enable forms of teaching and spaces for learning mathematics to possess an ecological sensibility?*

Jennifer sat. Keeping watch. But this time as she looked on she saw Stigler and Hiebert's question[5] from an opposite side.[6] Now from within an ecological mind-space, she realized that breaking out of one's conventions was not simply a matter of choice. Walking off the trail for Jennifer had proven to be an ongoing and challenging task of re-rooting the learning space in her classroom.

In teaching mathematics, she often heard that "good" teachers "go with the flow." In taking an ecological view however, Jennifer now understood responsive teaching to be much more than just going with the flow of the classroom and theoretical system or systems.

> "Teaching responsively as praxis involves considering how our teaching contributes to such a flow – conscious or not," she said.
>
> "So as well as participating in the practice of teaching mathematics, it's about having a concern for and need to make sense of one's participation in it. Thinking about mathematics teaching in this way allows for the emergence of deeper understandings about what it means to teach mathematics as well as why we might teach mathematics in the first place. A subtle yet fundamental difference, for sure," I said.
>
> "And it's a difference that I've come to realize, distinguishes this view in a radical and relevant way – because of the kind of examination that an ecological perspective calls for." "What's more," Jennifer said, "is how this difference arises from an implicit awareness and sense of responsibility that comes with knowing that every action or presence holds some sort of potential impact."
>
> "Impact that could ultimately affect the ecology or the oikos[7] of the classroom, right? " I said.
>
> "Yes, and beyond, " said Jennifer.
>
> "Oh right, of course. This was one of the points that came up in our first conversation about classroom mathematics, ecology, and the world as inseparable – everyone and everything is interconnected and therefore, interdependent. So, that's it?" I asked.
>
> Jennifer laughed. "There you go again, trying to simplify the complex!"

Stubbornly refusing my desire to tie things up, she smiled at me and offered Gary Snyder's comment, as what she called, an open-closure:[8]

> There is nothing like stepping away from the road and heading into a new part of the watershed. Not for the sake of newness, but for the sense of coming home to our whole terrain. 'Off the trail' is another name for the Way, and sauntering off the trail is the practice of the wild. That is also where – paradoxically – we do our best work. But we need paths and trails and will always be maintaining them. You must first be on the path, before you can turn and walk into the wild.[9]

"Let's take this," she said, "and set it side by side with Stigler and Hiebert's question and the issues that have surfaced regarding the need for an ecological sense of place in the mathematics classroom."

Together, we reflected on the question, revisiting past conversations about the nature of theory and practice in mathematics teaching and the manners in which the two co-exist and co-evolve as cognizing domains. Our return to these topics highlighted how assumptions, including those ideas that might be considered to be irrelevant to the teaching of mathematics – for example, conceptions of time, profoundly affected how Jennifer's teaching and the children's learning arose. It was obvious now that caring for students' learning in the mathematics classroom had more to do with the class connecting and engaging with place than it did with focusing directly on breaking out of and discarding conventions.

"In the context of Snyder's remarks, valuing place and how place relates to the teaching of mathematics emphasizes the importance of being aware of the paths that we teachers unthinkingly walk," said Jennifer.

"By paths," I said, "you mean, our assumed, embodied, and metaphorical ways of knowing, acting, and being, right?

"Exactly," said Jennifer.

"Okay, but if a path only exists in its enactment, how can there ever be such a thing as an established path or trail, like the ones that Snyder refers to?" I asked.

"Well, there can't be," Jennifer said, "or at least not in the sense that a path exists in the absence of our enactions and is some thing on which we walk time and time again. However, a path could be maintained or even well worn as a result of our recurrent enactments of it."

"Okay... yes, the metaphor makes sense then and, if you think about it with the idea of drift,[10]" I said.

"And what teaching as place-making implies is an engaged inquiry into the possibilities and impossibilities that our paths create for teaching and learning mathematics," said Jennifer.

> "So the point isn't for us to notice the conventions in our practices of teaching mathematics so that we may then regard them as redundant or unnecessary," I said.
>
> "That's right," said Jennifer, "it's about revealing how our manners of knowing, conventional or not, specify taken for granted ways of acting and being in the classroom which in turn, engender particular kinds of mathematics."
>
> " Because once revealed, we can take up the project of questioning and evaluating the sense of place and spaces that are part of the mathematics classroom like you did when you realized that your teaching reproduced mechanistic patterns. The fact that the path you discovered you were walking was a well worn or conventional one wasn't the point was it? It was the understanding of how the path you walked affected or as you described, *languaged* mathematics teaching and learning to be ecologically incoherent – linear, hierarchical, static and deterministic... I guess you could also say that re-searching your teaching in a systemic way produced the prompt and provided the means for you to lay down a different path – a path that allowed you to imagine new metaphors that now language in a radically changed manner, how you conceptualize and teach mathematics," I said.
>
> "Yes, it did. And mathematics teaching is all of this," said Jennifer, "but still, that's not all it is."
>
> "What do you mean?" I asked.
>
> "Well, mathematics teaching which possesses an ecological sensibility also provokes an awakening for that which is yet unknown. This kind of awareness *re-cognizes*[11] mathematical environments, mathematics, teaching, and learning to be co-emergent events, events that can only be brought forth as they occur moment-to-moment," said Jennifer.

As I reflected on what Jennifer had just said, I found myself tracing back through the lessons in which she worked to create an ecological sense of place in her classroom. Here, certain features about her teaching were evident: Scripting unscripted lessons,[12] mapping out potential as opposed to certain mathematics as part of her lesson preparation, distinguishing mathematics as both a residue and a source for learning,[13] making familiar mathematics unfamiliar through recursive teaching structures[14] and, attending to the class' individual and collective work.[15] While all of these features were integral to Jennifer's teaching and she purposefully used them to define the learning space, there also was not a lesson in which she or her students ever knew exactly what the end result would be or, even in a particular instance, what would happen next. Rather, it was in Jennifer and her students' moment-to-moment activity that the emergence and articulation of the class' mathematics were realized. These practices then, were deliberate attempts to not fix, separate, or predetermine what happened in the mathematics classroom but to embrace the simultaneous interdependence that she spoke of. Akin to Snyder's description of what it is like to step away from the road, it was this other path off the trail that Jennifer originated in walking which enabled teaching, learning, and mathematics to evolve as a dynamic whole and in a real sense, made it possible for her to enact what she intended teaching and learning mathematics to be.

Sitting with Jennifer within this ecological mind-space, I understood a little better and appreciated more, her reason for the open-closure of this conversation.

> "Teaching mathematics is far from simple," I said, "and minding where children's mathematics grow is not some*thing* that's answered once and for all. It's a genuine way of being that requires the persistent attention and responsive consideration of the teacher... for the particular and connected instances that define the mathematics classroom as place."

As we brought our conversation to a close, it also curiously, remained open. Jennifer explained that even though we may be tempted to move off of our conventional trails of teaching mathematics and think that our desire to do so will change our direction, it is when we become mindfully aware of the taken for grantedness that brings ease to the paths we walk that we may consider which paths might actually move us into the open terrain.

> "Only then," said Jennifer, "can thoughtful decisions be made about which paths we continue to walk and when it is time to
>
> l a y d o w n n e w o n e s...

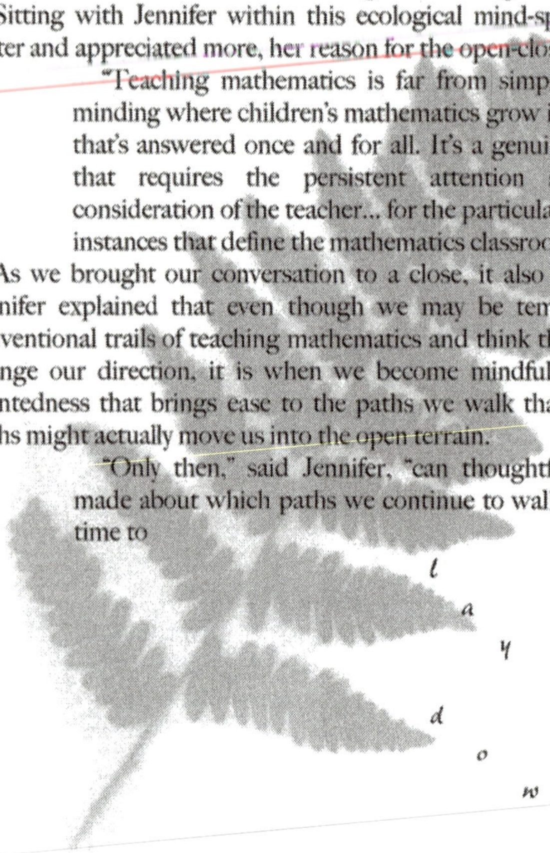

NOTES

Seamlessly and profoundly shaping my experiences, teaching, and research is the literature with which and through which I engaged during this study. What follows, is a collection of notes and an annotated bibliography of sorts. This set of references has been sequenced according to the book's physical front to back structure and by the conceptual clusters of the book. The notes are ordered and correspond with the text of specific compositions in the book that have the same superscript numerals.

Together, these references serve different purposes for the reader. The numerals in the text mark out particular authors and the work(s) from which their exact words were drawn, clarify certain ideas, or alert the reader to other places in the book that might be read next. The annotated bibliography highlights the specific ideas discussed in the book and references particular authors and works that may be of further interest to the reader.

ACKNOWLEDGMENTS (p. xi - xii)

1 Learning And The World We Want conference, "Untying a Dreamcatcher: Teaching Students of Aboriginal Inheritance" (a three act play), Victoria, BC, Canada, October 2003. Co-presenters: D. Blades and A. Oberg.

2 Presentation to the Faculty of Education in the *Conversations* series, "Untying a Dreamcatcher: Teaching Students of Aboriginal Inheritance" (a four act play), Victoria, Canada, April 2008. Co-presenters: D. Blades and A. Oberg.

3 A different version of the play was performed at the conference, Narrative Matters: Storying Our World Conference, "Untying a Dreamcatcher: Teaching Students of Aboriginal Inheritance" (a four act play), Toronto, ON, Canada, May 2008. Co-presenters: D. Blades and A. Oberg.

4 Oberg, A., Blades, D. & Thom, J. S. (2007). Untying a dreamcatcher: Coming to understand possibilities for teaching students of aboriginal inheritance. *Educational Studies, 42*(2), 111-139.

5 Thom, J. S. & Pirie, S. E. B., (2001). Thinking through ecological metaphors in mathematics education. In S. Gunn & A. Begg (Eds.), *Mind, Body, and Society: Emerging Understandings of Knowing and Learning*. Melbourne, Australia: University of Melbourne, pp. 45-52.

6 24th North American Chapter of the International Group for the Psychology of Mathematics Education, "Mathematics Teaching and Metaphorical Mattering", Athens, United States, October 2002.

7 Complexity Science and Educational Research Conference, "Distinguishing Difference in the Mathematics Classroom", Loranger, United States, November 2005.

8 National Council of Teachers of Mathematics Regional Conference, "Connecting Children with *Their* Mathematics", Regina, SK, Canada, October 2002.

9 2nd Meeting of Enactivism and Post-structuralism: Implications for Development of Educators, "Enactivism, Collaborative Communities, and Teacher Development", Melbourne, Australia, November 2002.

10 Thom, J. S. (2003). Beyond individual classrooms: Opening collaborative spaces for teaching and learning mathematics. *The Numerator, 2(2)*, 13-16.
11 American Education Research Association Annual Meeting, "What do you know about 72? Can We See the Connections Between the Images and the Formalizing?", Seattle, United States, April 2001.
12 Thom, J. S. & Pirie, S. E. B. (2006). Looking at the complexity of two young children's understanding of number. *Journal of Mathematical Behavior, 25*(3), 185-195.

WE ARE CONNECTED TO THIS EARTH (p. 1 - 3)

1 F. Capra, 1996, p. 7. A critique of western traditions of linear, mechanistic, and reductionist thought. The author reviews holistic ontological and epistemological orientations informed by developments in biology, mathematics, physics, psychology, ecology, and cybernetics. See also, F. Capra (2002) in which the author provides the reader with a general study of social, cultural, and global issues from perspectives grounded in complexity and systems thinking.
2 N. Evernden, 1993, p. 40-41.
3 D. Suzuki, 1997, p. 199.
4 S. Kotagama, 1993, p. 120.
5 As cited in S. Kotagama, 1993, p. 120-121.

For general discussions that contrast between shallow and deep ecological orientations, see A. Naess (1985), W. Fox (1990), and D. Orr (1992). For further discussions about the paradigmatic differences that characterize deep ecological sensibilities, see for example, A. Naess (1988), D. Orr (1994), D. Suzuki & H. Dressel (1999), and D. Suzuki & D. Taylor (2009).

SIGNIFICANCES IN THE FLOW OF INTERACTING (p. 5 - 10)

1 P. Bunnell, 2000, p. 31.
2 D. Abram, 1996, p. 15. Abram explores the meanings that arise through experiences with the more-than-human world. See also, D. Abram (2010).
3 M. C. Bateson, 1996, p.78.

D. Worster (1990) provides a historical account concerning the evolution and associated meanings of the word, *ecology* in his book, *Nature's economy: A history of ecological ideas*, see p. 191-192. As it was my desire to be ecologically responsive as a teacher, it was the notion of responding in ways which emerge from a natural will or sense of care (as opposed to prescribed impositions of 'oughts') that prompted me to consider my conceptualization of mathematics teaching and learning in the classroom. For discussions on this, see A. Naess, (1986, 1996, 2008) as well as F. Varela (1999).

SPACE WANTED (p. 11)

1 J. Stigler & J. Hiebert, 1997, p. 14.
2 See M. C. Bateson (1996).
3 G. Bateson, 1980, p. 9.
4 See T. Kieren, S. Pirie, & L. Gordon Calvert (1999) for a theoretical discussion and classroom episodes that illuminate mathematical understanding as inherently personal, contextual, and co-emergent.

ENCIRCLING OUR PERCEPTIONS (p. 13 - 29)

1 Invocative interventions are those that engender folding back – a recursive form of revisiting to earlier layers of understanding.
2 Provocative interventions are those that enable learners to develop new and deeper understandings.
3 Emphasis added, B. Davis et al., 2000, p. 64.
4 R. Lewontin, 1983, p. 75-76.
5 G. Bateson, 1980, p. 9.
6 H. Maturana, 1988b, p. 46.
7 F. Varela et al., 1996, p. 205.
8 H. Maturana, 1988b, p. 41.
9 N. Postman 1995, p. 181.
10 F. Capra, 1996, p. 7.

For descriptions and explanations of H. Maturana and F. Varela's theory of autopoiesis, see H. Maturana (1987) and H. Maturana & F. Varela (1980). For a detailed description of structural determinism, see H. Maturana, (1988b), p. 36. For an explanation about the process of structural coupling, see H. Maturana (1988a). For the definition of natural drift, see H. Maturana & F. Varela (1987). For another similar discussion on the dialectic process of knowing and the environment as it applies to mathematics, see L. Steffe and R. Tzur (1994). See H. Maturana (1991, 2001) for the dynamical nature of social systems. In H. Maturana's (1997a, 1997b) papers, he describes the notion of languaging within collectives. See H. Maturana (1998) for an account of the role that cultural drift plays in the emergence and evolution of cultural systems. Each in their own ways, C. Bowers (2003), F. Capra (1996), A. Naess (1988) and D. Orr (1992) provide discussions on how cultural ways of being define our perceptions and existence in the world. See H. Maturana & F. Varela (1987) for their examination concerning the circular nature of cognition. See G. Bateson (1972) for his interpretation and distinction between the notion of map and territory and also, see C. Bowers (1997) for a discussion on map and territory in relation to G. Bateson's definition. Both G. Bateson and C. Bowers base their conceptualizations on A. Korzybski's previous work in general semantics, specifically, his ideas about map and territory. For example, see A. Korzybski (1931).

HELP WANTED (p. 31 - 44)

1 J. Stigler & J. Hiebert, 1997, p. 14.
2 See M. C. Bateson (1996).
3 G. Bateson, 1980, p. 9.
4 See T. Kieren, S. Pirie, & L. Gordon Calvert (1999) for a theoretical discussion and classroom episodes that illuminate mathematical understanding as inherently personal, contextual, and co-emergent.
5 J. Stigler & J. Hiebert, 1997, p. 14.
6 See p. 13-29 for a discussion on how we exist as members of cognitive beings within interdependent yet distinct systems.
7 See p. 25-27 for an explanation of cognitive circles and cultural maps.
8 K. Basso, 1996, p. xiii.
9 B. Davis, 1996, p. 132.
10 D. Jardine, 1994, p. 109.
11 M. C. Bateson, 1994, p. 31.

See A. Camus (1955) for how sense of place is brought forth and exists in being through our actions. The sense of playfulness that is evoked through the image of adventuring resonated with my intentions as a mathematics teacher. Such teaching and learning are found within the schools of Reggio Emilia, Italy.

There is a wealth of literature about this educational approach and examples of how the ideas of Reggio Emilia preschool and primary schools have been implemented in a variety of learning contexts around the world. For detailed descriptions and narratives of the history, guiding philosophies, and exemplars from Reggio Emilia classrooms, see C. Edwards et al. (1993).

TRADITIONS, TENSIONS, AND TRANSITIONS (p. 45 - 72)

1 J. Stigler & J. Hiebert, 1997, p. 14.
2 See M. C. Bateson (1996).
3 G. Bateson, 1980, p. 9.
4 See T. Kieren, Pirie, & L. Gordon Calvert (1999) for a theoretical discussion and classroom episodes that illuminate mathematical understanding as inherently personal, contextual, and co-emergent.
5 B. Davis et al., 2000, p. 41.
6 J. Scieszka, and Smith, 1995, p. 27.
7 M. C. Bateson, 1994, p. 14.
8 H. Maturana, 1997b, p. 5.
9 See "Vacancy" advertisement, p. 45.
10 W. Berry, 1983, p. 34.
11 For a discussion about place, see p. 34-35.
12 NCTM (2000).
13 As discussed by A. Baroody (1993).
14 For the meaning of the word, oikos, see p. 8. For how the meaning of the word informed my conceptualization of classroom mathematics, see, p. 9.
15 Emphasis added, D. Jardine, 1999, p. 35.
16 Emphasis added, as cited in D. Jardine, 1999, p. 35.
17 Paraphrased from M. Heidegger (1962). See p. 188-203.
18 Emphasis added, G. Lakoff & R. Núñez, 2000, p. 39.
19 A. Korzybski, 1931, p. 750.
20 Emphasis added, N. Postman, 1995, p. 181.
21 For examples of emergent mathematics, see p. 146-157, 224-229, 313-317.
22 For examples of integrative mathematics, see p. 245-263, p. 323-339, and p. 341-345.
23 For examples of individual-within-collective mathematics, see p. 151-157 and 269-284.
24 D. Jardine, 1999, p. 37.

Mrs. Fibonacci is the teacher in the children's book, *Math Curse* written by J. Scieszka and illustrated by L. Smith (1995). It was through G. Bateson's idea of "the pattern which connects" that I focused my examination of the planned and lived aspects of my mathematics program. For G. Bateson's discussion of this, see G. Bateson (1980). T. Aoki and B. Uhrmacher in different yet compatible ways, both examine how a teacher's past experiences shape present teaching and learning contexts and become what Aoki calls the "lived curriculum" and what Uhrmacher refers to as the "curriculum shadow". For these discussions, see T. Aoki (1993) and B. Uhrmacher (1997). Several works by A. Sfard, D. Abram, F. Capra, M. C. Bateson, M. van Manen, D. Jardine, D. Orr, H. Maturana & F. Varela, G. Lakoff & M. Johnson are mentioned in other parts of this book. Additional examples of these and other authors whose thinking also influenced my thinking included: F. Capra (1993), M. C. Bateson (2000), M. van Manen (1991), D. Jardine (2000), and G. Lakoff & M. Johnson (1980, 1999). Traditionally and currently in British Columbia, each subject area for grades K-12 is published by the Ministry as a separate curriculum document. For example, see Ministry of Education (2007).

METAPHORICAL MATTERING OF MATHEMATICS (p. 73 - 76)

1 K. Devlin, 2000, p. 92.

For discussions of western practices that define the nature and role that mathematics plays in education and society at large, see B. Davis (1995), K. Devlin (1994), and D. Wheeler (1967).

THEORETICAL PORTRAITS OF MATHEMATICAL UNDERSTANDING (p. 77 - 135)

Figures 9 and 11 were reprinted with permission from the *Journal for Research in Mathematics Education*, copyright 1996 by the NCTM. All rights reserved. Figures 13-18 were reprinted with permission from *Handbook of research on mathematics teaching and learning: A project of the national council of teachers of mathematics*, copyright 1992 by the NCTM. All rights reserved. All references to the literature are cited in the compositions except for the one that appears in the text portrait on p.134:

1 H. Maturana & F. Varela, 1987, p. 27.

NEW FURNITURE (p. 137 - 142)

1 For different interpretations of mathematics, see p. 73-76.
2 For a review and discussion of key theoretical frames used in interpreting mathematical understanding and implications for mathematics teaching and learning, see p. 77-135.
3 B. Davis et al., 2000, pg. 62.
4 B. Davis et al., 2000, pg. 62.
5 B. Davis et al., 2000, pg. 62.
6 F. Varela et al., 1996, p. 205.
7 For the meaning of the word, oikos, see p. 8.
8 See also p. 269-284 for examples of children's shape-shifting of mathematical ideas.
9 This is a further consideration of what was initially prompted by R. Davis' (1992) thinking about mathematics as residue.

See G. Bateson (1972) for his interpretation and distinction between the notion of map and territory that draws on A. Korzybski's previous work (e.g., see A. Korzybski, 1931) in general semantics around these very ideas.

LANGUAGING EMERGENT MATHEMATICS (p. 143 - 161)

1 NCTM (2000).
2 As discussed by A. Baroody (1993).
3 For a discussion on "language" see p. 22-28.
4 See NCTM, 2000, p. 30-31.
5 See p. 125-127 for A. Sfard's conception of mathematical thinking as communication.
6 Both terms, source and residue are described from an ecological perspective on p. 142 and p. 159-160. Also, see J. Thom & S. Pirie (2001) and J. Thom & S. Pirie (2002).
7 For a further discussion on specific individual and collective aspects of this lesson's mathematical residue, see p. 266-268.
8 For a discussion of this other space, see p. 54-56.
9 See p. 64-66 for the mechanistic metaphors and p. 140-142 and p. 170 for the ecological metaphors that structured my thinking.

The story of *Even Steven and Odd Todd* was written by K. Cristaldi and illustrated by H. Morehouse (1996). During my reflection on the lesson of even and odd numbers, I had a conversation with S. Pirie in which we discussed at length, the meanings she holds for such so called, "liberating constraints". See B. Davis, et al. (2000) for their elaboration on this idea of liberating constraints and how they describe these as a teacher's efforts to strike a "balance between freedom and restraint that creates conditions for learning and creativity" (p. 87). For the five process standards, see NCTM (2000) and in particular, see p. 29 for full descriptions of the council's standards for school mathematics preK - 12. See T. Roszak (1994) for his discussion of and differentiation between information and ideas. For a description of domain as is applies to cognition and the role that consensuality plays in giving rise to such domains, see H. Maturana (1988b).

ATMOSPHERE AND RE-COGNIZING DIFFERENCES (p. 163 - 171)

1 van Manen, 1986, p. 32.
2 van Manen, 1986, p.31.
3 C. Bowers and D. Flinders, 1991, p. 11.
4 For the meaning of the word, oikos, see p. 8. For how the meaning of the word informed my conceptualization of classroom mathematics, see, p. 9.
5 Of course, this also includes the taken-for-granted language and languaging of society and particularly, how this has influenced the students.
6 See p. 140-142 and p. 170.
7 F. Varela et al., 1996, p. 205.
8 A diagram based on S. Pirie & T. Kieren's (1994a) model for the growth of mathematical understanding.

EMBEDDING AN ECOLOGICAL SENSE OF PLACE FOR MATHEMATICS IN THE CLASSROOM (p. 173 – 180 and 199 - 359)

1 H. Maturana, 1997b, p. 5.

ATTENDING TO THE PHYSICAL SPACE IN THE CLASSROOM (p. 175 - 180)

1 C. Rinaldi, 1990 as cited in C. Edwards et al., 1993, p. 137.
2 These are interlocking polygons used for building 2D and 3D figures.

CONCEPTUALIZATIONS OF TIME (p. 181 – 198)

1 J. Greenman, 1988, p. 5.
2 Emphasis added, I. Newton, 1687 as cited in A. Koyré, 1958, p. 161-162.
3 D. Abram, 1996.
4 Aristotle, 1969, p. 81.
5 T. C. McLuhan, 1987, p. 32-33.
6 J. Kabat-Zinn, 1994, p. 4.
7 An excerpt from H. Maturana's (2001) lecture.
8 An excerpt from H. Maturana's (2001) lecture.
9 M. Heidegger, 1924, p. 14E.
10 M. Heidegger, 1924, p. 22E.

For discussions on the complex and recursive nature of how time is experienced and perceived, see D. Abram (1996), C. Bowers (1997), and H. Maturana (1995, 2001).

ENACTING RE-CONCEPTUALIZATIONS OF TIME (p. 199 - 235)

1 See p. 140-142 and p. 170 for my new ecological metaphors.
2 For a discussion of the differences between complicated and complex events, see p. 138-139.
3 This idea of an inside core parallels the Pirie-Kieren (1994b) dynamical model of growth of mathematical understanding which defines primitive knowing as being all the knowledge that a child or group of children bring to the particular mathematics and from which all new understandings develop. It is this embodied mathematical knowledge that I assumed my students possessed upon entering the two-year program.
4 These are the learning goals that are prescribed by the BC Ministry of Education and published in the mathematics integrated resource package for grades K-7.
5 For example, resource material and educational literature published by the National Council for Teachers of Mathematics (NCTM) and British Columbia Association for Mathematics Teachers (BCAMT).
6 In the same manner as the themes in this program, the mathematical investigations were open ended in structure. Emphasis was placed on the students' active engagement in problem solving and problem posing within nonroutine settings. For example, see S. Brown and M. Walter (1983), N. Gonzales (1994), R. Lesh (1981), and M. Walter & S. Brown (1993).
7 See p. 40 for M.C. Bateson's description.
8 See p. 124-134 for a discussion on conceptions of mathematical understanding as viewed from an enactive perspective.
9 See p. 16 for a discussion on structural coupling.
10 See p. 128-129 for a model and description of the Pirie-Kieren theory.
11 When I identified these four qualities of recursion, they unexpectedly all began with the letter, R! The resulting alliteration is not only serendipitous in that these 4Rs contrast with the traditional 3Rs, reading, writing, and arithmetic, but they also complement the curricular 4Rs, richness, relations, rigor, and in particular, recursion as explicated by Doll Jr. (1993).

For examples of the teaching and learning within schools of Reggio Emilia, see R. New (1993). *Notes on a Triangle* is a film produced by the National Film Board (1969). See M. C. Bateson (1994) for her description of learning as similar to moving along a Möbius strip. See Kieren & Pirie (1991) for explication of the qualities that characterize conceptual understanding as a recursive event.

MAKING (THREE) SPACE(S) FOR RECURSION (p. 209 - 212)

1 This project was adapted from the activity, *Find the Secret Number*. See W. Liedtke, 1983, p. 89.
2 See p. 129 for a definition of image (making and having).
3 See p. 129 for a definition of folding back.
4 See p. 129 for a definition of property noticing.

A SNOWFLAKE OF A DIFFERENT KIND (p. 213 - 221)

1 The Koch snowflake is a fractal curve based on the Koch curve and named after the mathematician, Helge von Koch. The Koch snowflake possesses infinite perimeter while at the same time, a finite area.
2 This idea of keeping watch is discussed on p. 57.

RE-VIEWING AND SEEING DIFFERENTLY (p. 223 - 229)

Notes on a Triangle is a film produced by the National Film Board (1969).

RECURSION AS RELATIONS (p. 231 - 234)

1 See p. 129 for a definition of property noticing.

Dance Squared is a film produced by the National Film Board (1961).

TWO AND A HALF YEARS LATER... (p. 235)

For a phenomenological examination of time as a recursive event, see M. Merleau-Ponty (1968). For descriptions and articulations of time as an all-at-once phenomenon, see M. Heidegger (1972) and H. Maturana (1995). See M. Heidegger (1962) for an explication of the inseparability of time and space.

CREATING PATTERNS WHICH CONNECT (p. 237 - 302)

1 G. Bate son, 1980, p. 18.
2 M. van Manen, 1986, p. 44.
3 G. Bate son, 1980, p. 9.

INTERACTIONAL SPACES FOR MATHEMATICS (p. 243 - 302)

UNDERSTANDINGS THAT SPEAK TO ONE ANOTHER (p. 245 - 264)

1 M. C. Bateson, 1994, p. 14.
2 See p. 129 for a definition of image.
3 See p. 129 for a definition of folding back.

FREE IDEAS AND CONNECTING AS SHAPE-SHIFTING (p. 265 - 284)

1 M. C. Bateson, 1994, p. 14.
2 R. New, 1993, p. 219.
3 For a discussion on collective understanding, see p. 132-133.
4 See p. 151-157 and p. 269-284 for classroom examples of the individual and collective qualities of the class' understanding.
5 This graph was adapted from AIMS (Activities Integrating Mathematics and Science), 1987, p. 70. Reprinted by permission of the AIMS Education Foundation from *Primarily Bears*, copyright 1987 AIMS Education Foundation. All rights reserved. Visit http://www.aimsedu.org for more information.
6 This metaphor first emerged while the students were investigating even and odd numbers. See p. 153.

OPENING MATHEMATICAL SPACES OF THEIR OWN (p. 285 – 302)

1 See p. 129 for definition of folding back.
2 See p. 129 for a definition of image (making and having).
3 See p. 129 for a definition of property noticing.
4 In hindsight, had I asked him to explain whether 33, 55, 77, or 75 was even or odd, this might have given me the insight into interpreting his understanding here.

MATHEMATICS BEYOND THE CLASSROOM (p. 303 – 320)

For the epsistemological orientation and meaning of G. Bateson's idea of the patterns which connect, see G. Bateson, 1980, p. 9. See H. Gadamer (1989) for insight into meaning making as arising through genuine conversation. *Dance Squared* is a film produced by the National Film Board (1961). *Selina and the Bear Paw Quilt* is children's book written by B. Smucker and illustrated by J. Wilson (1995).

BEYOND SPACES OF INTERACTION (p. 321 - 345)

1 D. Jardine, 1994, p. 112.

CONNECTING US TO IT (p. 341 - 345)

1 For a discussion on this idea of free ideas, see p. 266.

In his books, D. Abram (1996, 2010) questions, engages, and reveals a most provocative view of the more-than human world. For the epistemological orientation and meaning of G. Bateson's idea of the patterns which connect, see G. Bateson, 1980, p. 9. See M. van Manen, 1986, p. 44 for a phenomenological discussion of meaning as relationship.

SPACES FOR UNPREDICTABLE MATHEMATICS (p. 347 - 359)

1 R. Fulghum, 1989, p. 81-82.
2 H. Burkhardt, 1988, p. 18.
3 H. Burkhardt, 1988, p. 18.
4 See p. 142 for a description, p. 159-160, for a discussion, and p. 150-155 for examples of mathematics as residue.
5 See p. 142 for a description, p. 159-160, for a discussion, and 156-157 for examples of mathematics as source.
6 R. Fulghum, 1989, p. 83.

OFF THE TRAIL (p. 361 – 366)

1 J. Stigler & J. Hiebert, 1997, p. 14.
2 See M. C. Bateson (1996).
3 G. Bateson, 1980, p. 9.
4 See T. Kieren, S. Pirie, & L. Gordon Calvert (1999) for a theoretical discussion and classroom episodes that illuminate mathematical understanding as inherently personal, contextual, and co-emergent.

5 J. Stigler and J. Hiebert, 1997, p. 14.

6 M. C. Bateson, 1994, p. 31.

7 For the definition of oikos, see p. 8.

8 See p. 349 and p. 350-351. See also, p. 142, 156-157, and p. 159-160 for related discussions and examples of mathematics as source.

9 G. Snyder, 1990, p. 154.

10 For a description of drift as recurrent interactions, see p. 19.

11 Here as implied in the discussion of re-cognizing the notion of difference on p. 169, the word, *re-cognize*, from the Latin word recognoscere, should not be interpreted with its usual meaning, to "identify as already known" but rather, with the word's secondary meaning which is to "know again" as in, to know anew. To re-cognize classroom mathematics in this sense then involves the teacher recursively *again and again* (RE) conceptualizing the mathematical environment(s), mathematics, teaching, and learning *together* (CO) in necessarily dynamic, collective, and co-determinant manners in order to *know* (GNOSCERE).

12 See p. 147, 151-154, and 157 for examples of unscripted lessons.

13 See p. 150-155 for examples of mathematics as residue and p. 156-157 for examples of mathematics as source.

14 See p. 209-234 for examples of recursive lesson structures.

15 See p. 151-157 and p. 269-284 for classroom examples and p. 266-268 for a discussion of individual and collective mathematics in the classroom.

References

Abram, D. (1996). *The spell of the sensuous: Perception and language in a more-than-human world.* New York: Vintage Books.
Abram, D. (2010). *Becoming animal: An earthly cosmology.* New York: Pantheon.
AIMS (Activities Integrating Mathematics and Science). *Primarily bears.* Fresno, CA: AIMS Educational Foundation.
Aoki, T. T. (1993). Legitimating lived curriculum: Towards a curricular landscape of multiplicity. *Journal of Curriculum and Supervision, 8*(3), 255-268.
Aristotle (1969). *Physics.* (H. G. Apostle, Trans.) (book IV). Bloomington, IN: Indiana University.
Bachelard, G. (1975). *Le nouvel esprit scientifique.* Paris: Universitaires de France.
Bachelard, G. (1983). *La formation de l'esprit scientifiqueu.* Paris: Universitaires de France.
Baroody, A. J. (1993). *Problem solving, reasoning, and communicating (K-8): Helping children think mathematically.* New York: Macmillan.
Basso, K. H. (1996). *Wisdom sits in places: Landscape and language among the western apache.* Albuquerque, NM: University of New Mexico.
Bateson, G. (1972). *Steps to an ecology of mind.* New York: Ballantine.
Bateson, G. (1980). *Mind and nature: A necessary unity.* New York: Dutton.
Bateson, M. C. (1994). *Peripheral visions: Learning along the way.* New York: HarperCollins.
Bateson, M. C. (1996). Democracy, Ecology, and Participation. In Roger Soder (Ed.) *Democracy, Education, and the Schools.* (pp. 69-86). San Francisco, CA: Jossey-Bass.
Bateson, M. C., (2000). *Full circles, overlapping lives: Culture and generation in transition.* New York, NY: Ballantine.
Bergeron, J. C., Herscovics, N., & Bergeron, A. (1986). The kindergartner's symbolization of numbers. In G. Lappan & R. Even (Eds.), *Proceedings of the 8th Annual Meeting of the North American Chapter of the International Group for the Psychology of Mathematics Education* (pp. 35-41). East Lancing, MI: Michigan State University.
Berliner, P. (1994). *Thinking in jazz: The infinite art of improvisation.* Chicago, ILL: University of Chicago.
Berry, W. (1983). *Standing by words.* San Francisco, CA: North Point.
Beth, E. W. & Piaget, J. (1966). *Mathematical epistemology and Psychology.* Dordrecht, Holland: D. Reidel .
Biggs, J. B. & Collis, K. E. (1982). *Evaluating the quality of learning: The SOLO taxonomy (structure of the observed learning outcome).* New York: Academic.
Biggs, J.B. & Collis, K.F. (1991). Multimodal learning and the quality of intelligent behaviour. In H.Rowe (Ed.) *Intelligence: Reconceptualisation and measurement.* Hillsdale, NJ: Lawrence Erlbaum.
Bishop, A. J., (1986). What are some obstacles to learning geometry? *Studies in Mathematics Education (UNESCO), 5,* 142-159.
Blake, A. J. D. (1978). Level of intellectual development of teacher education students. *Australian Journal of Education, 22*(1), 89-90.
Bowers, C. A. (1997). *The culture of denial: Why the environmental movement needs a strategy for reforming universities and public schools.* Albany, NY: State University of New York.

Bowers, C. A. (2003). *The false promises of constructivist theories of learning: A multicultural and ecological perspective.* Unpublished book manuscript.
Bowers, C. A. & Flinders, D. J. (1991) *Culturally responsive teaching and supervision: A handbook for staff development.* New York, NY and London, England: Teachers College.
Bowers, J. S. & Nickerson, S. (2001). Identifying cyclic patterns of interaction to study individual and collective learning. *Mathematical Thinking and Learning 3*(1), 1-28.
Brown, S. I., & Walter, M. I. (1983). *The art of problem posing.* Philadelphia, PA: Franklin Institute.
Brownwell, W. A. (1935). Psychological considerations in the learning and teaching of arithmetic. In W. D. Reeve (Ed.), *The teaching of arithmetic: Tenth yearbook of the National Council of Teachers of Mathematics,* (pp. 1031). New York: Columbia University, Teachers College.
Bunnell, P. (2000). *Understanding as a path.* Vancouver, BC: LifeWorks.
Burkhardt, H. (1988). Teaching problem solving. In H. Burkhardt, S. Groves, A. Schoenfeld, & K. Stacey (Eds.), *Problem solving– A worldview. Proceedings of the problem solving theme group, ICME 5* (pp. 17-42). Nottingham, England: Shell Centre.
Camus, A. (1955). *Noces suivi de l'eté.* Paris: Gallimard.
Capra, F. (1993). *Guide to ecological literacy.* Berkely: The Elmwood Institute.
Capra, F. (1996). *The web of life: A new scientific understanding of living systems.* New York: Anchor Books.
Capra, F. (2002). *The hidden connections: integrating the biological, cognitive, and social dimensions of life into a science of sustainability.* New York: Doubleday.
Carpenter, T. P., Hiebert, J., & Moser, J. M., (1983). The effect of instruction on children's solutions of addition and subtraction word problems. *Educational Studies in Mathematics, 14*(2), 55-72.
Carraher, T. N., Carraher, D. W., & Schliemann, A. D. (1987). Written and oral mathematics. *Journal for Research in Mathematics Education, 18*(2), 83-87.
Clements, K. (1982a). Visual imagery and school mathematics (part 1). *For the Learning of Mathematics, 2*(2), 2-39.
Clements, K. (1982b). Visual imagery and school mathematics (concluded). *For the Learning of Mathematics, 2*(3), 33-39.
Cobb, P. (1999). Individual and collective mathematical development: The case of statistical data analysis. *Mathematical Thinking and Learning, 1*(1), 5-43.
Cottrill, J., Dubinsky, E., Nichols, D., Schwingendorf, K., Thomas, K., & Vidakovic, D. (1996). Understanding the limit concept: Beginning with a coordinated process scheme. *Journal of Mathematical Behavior 15*(2), 167-192.
Cristaldi, K. (1996). *Even Steven and odd Todd.* New York: Scholastic.
Davis, B. (1995). Why teach mathematics? Mathematics education and enactivist theory. *For the Learning of Mathematics, 15*(2), 2-9.
Davis, B. (1996). *Teaching mathematics. Toward a sound alternative.* New York, NY & London, England: Garland.
Davis, B. & Simmt, E., (2003). Understanding learning systems: Mathematics education and complexity science. *Journal of Research in Mathematics Education 34*(2), 137-167.
Davis, B., & Sumara, D. J. (2000). Curriculum forms: on the assumed shapes of knowing and knowledge. *Journal of Curriculum Studies, 32*(6), 821-845.
Davis, B., Sumara, D. J., & Luce-Kapler, R. (2000). *Engaging minds: Learning and teaching in a complex world.* Mahwah, NJ & London, England: Lawrence Erlbaum.
Davis, R. B. (1984). *Learning mathematics: The cognitive science approach to mathematics education.* Norwood, NJ: Ablex.
Davis, R. B. (1992). Understanding "understanding". *Journal of Mathematical Behavior, 11*(3), 225-241.
Davis, R. B. & Vinner, S. (1986). The notion of limit: Some seemingly unavoidable misconceptions stages, *Journal of Mathematical Behavior, 5*(3), 281-301.
Devlin, K. (1994). *Mathematics: The science of patterns.* New York: Scientific American Library.
Devlin, K. (2000). *The maths gene: Why everyone has it, but most people don't use it.* London, England: Weidenfield & Nicolson.
Dewey, J. (1988). *Jak myslimy?* Warzawa: PWN. Polish translation of "how we think?" First published in USA, 1910.

Dionne, J. & Boukhssimi, D. (1988). Paper presented at working session, Montreal, QC.

Doll, Jr., W. (1993). *A post-modern perspective on curriculum.* New York: Teachers College.

Dubinsky, E. (1992). Reflective abstraction in advanced mathematical thinking. In David Tall (Ed.), *Advanced mathematical thinking.* Dordrecht: Kluwer.

Dubinsky, E., & Lewin, P. (1986). Reflective abstraction and mathematics education: The genetic decomposition of induction and compactness. *Journal of Mathematical Behavior, 5,* 55-92.

Durkin, K., & Shire, B. (1991). *Language in mathematical education: Research and practice.* Milton Keynes: Open University.

Edwards, C., Gandini, L., & Forman, G. (Eds.) (1993). *The hundred languages of children.* Norwood, NJ: Ablex.

Eisenberg, T., & Dreyfus, T. (1991). On the reluctance to visualize in mathematics. In W. Zimmermann & S. Cunningham (Eds.), *Visualization in teaching and learning mathematics* (pp. 25-38). MAA Notes #19.

Evernden, N. (1993). *The natural alien: Humankind and environment,* 2nd ed., Toronto, ON: University of Toronto Press.

Fox, W. (1990). *Toward a transpersonal ecology.* Boston: Shambhala.

Fulghum, R. (1989). *All I really need to know I learned in kindergarten: Uncommon thoughts on common things.* New York: Ballantine

Gadamer, H. G. (1989). *Truth and method* (J. Weinsheimer & D. G. Marshall, Trans.) (2nd rev. ed.). New York: Crossroads.

Goldenberg, E. P. (1991). See beauty in mathematics: Using fractal geometry to build a spirit of mathematical inquiry. In W. Zimmermann & S. Cunningham (Eds.), *Visualization in teaching and learning mathematics* (pp. 39-66). MAA Notes #19.

Gollwitzer, H. (1991). Visualization in differential equations. W. Zimmermann & S. Cunningham (Eds.), *Visualization in teaching an learning mathematics* (pp. 149-156). MAA Notes #19.

Gonzales, N. A. (1994). Problem posing: A neglected component in mathematics courses for prospective elementary and middle school teachers. *School Science and Mathematics, 94,* 78-84.

Gordon Calvert, L. (1999). *Mathematical conversations within the practice of mathematics.* Unpublished doctoral dissertation. University of Alberta, Edmonton, AB, Canada.

Gravemeijer, K., Cobb, P., Bowers, J., & Whitenack, J. (2000). Symbolizing, modeling, and instructional design. In P. Cobb, E. Yackel, & K. McClain (Eds.), *Symbolizing and communicating in mathematics classrooms* (pp. 225-273). Mahwah, NJ: Erlbaum.

Greenman, J. (1988) *Caring spaces, learning spaces: Children's environments that work.* Redmond, VA: Exchange.

Hallam, R. N. (1967). Logical thinking in history. *Educational Review, 19*(3), 183-202.

Hart, L. C. (1989). Some factors that impede or enhance performance in mathematical problem solving. *Journal for Research in Mathematics Education, 24*(2), 167-171.

Heidegger, M. (1924). *The Concept of Time* (W. McNeill, Trans.). Oxford: Blackwell.

Heidegger, M. (1962). *Being and time.* (J. Macquarrie & E. Robinson, Trans.). New York, NY: Harper.

Heidegger, M. (1972). *Time and being. On being and time.* (J. Stambaugh, Trans.). New York, NY: Harper & Row.

Hèraurd, B. (1988). Plan de thèse de doctorat, Université de Montréal.

Herscovics, N. (1989). The description and analysis of mathematical processes. In C. Maher, G. A. Goldin, & R. B. Davis (Eds.), *Proceedings of the 11th Annual meeting of the North American Chapter of the International Group for the Psychology of Mathematics Education* (pp. 9-36). Rutgers, NJ: Rutgers University.

Herscovics, N. & Bergeron, J. C. (1981). Psychological questions regarding a new model of understanding elementary school mathematics. In T. R. Post & M. P. Roberts (Eds.), *Proceedings of the 3rd Annual Meeting of the North American Chapter of the International Group for the Psychology of Mathematics Education* (pp. 69-76). Minneapolis, MN: University of Minnesota.

Herscovics, N. & Bergeron, J. C. (1982). A constructivist model of understanding. In S. Wagner (Ed), *Proceedings of the 4th Annual Meeting of the North American Chapter of the International Group for the Psychology of Mathematics Education* (pp. 28-35). Athens, GA: University of Georgia.

Herscovics, N. & Bergeron, J. C. (1983). Models of understanding. *Zentralblatt fur Didaktik der Mathematik, 15*(2), 75-83.

Herscovics, N. & Bergeron, J. C. (1984). A constructivist vs. a formalist approach in the teaching of mathematics. In B. Southwell et al. (Eds), *Proceedings of the 8th Annual Meeting of the International Group for the Psychology of Mathematics Education* (pp. 190-196). Sydney, Australia: University of Sydney.

Herscovics, N. & Bergeron, J. C. (1988a). An extended model of understanding. In M. J. Behr, C. B. Lacampagne, & M. M. Wheeler (Eds.), *Proceedings of the 10th Annual Meeting of the North American Chapter of the International Group for Psychology of Mathematics Education* (pp. 15-22). DeKalb, ILL: Northern Illinois University.

Herscovics, N. & Bergeron, J. C. (1988b). The kindergartners' understanding of the notion of rank. In A. Borbas (Ed.), *Proceedings of the 12th Annual Meeting of the International Group for the Psychology of Mathematics Education* (Vol. 1, pp. 385-392). Veszprèm, Hungary: Ferene, Genzqein, OOK.

Herscovics, H., Bergeron, J. C. & Bergeron, A. (1986a). Counting procedures used by kindergarten children. In the *Proceedings of the 10th Annual Meeting for the International Group for the Psychology of Mathematics Education* (pp. 43-48). London, England: University of London Institute of Education.

Herscovics, H., Bergeron, J. C. & Bergeron, A. (1986b). The kindergartner's perception of the invariance of number under various transformations. In G. Lappan & R. Even (Eds.), *Proceedings of the 8th Annual Meeting of the North American Chapter of the International Group for the Psychology of Mathematics Education* (pp. 28-34). East Lansing, MI: Michigan State University.

Hiebert, J. (1985). (Ed.) *Conceptual and procedural knowledge: The case of mathematics.* Hillsdale, NJ: Lawrence Erlbaum.

Hiebert, J. (1989). The struggle to link written symbols with understandings: An update. *Arithmetic Teacher, 36*(7), 38-44.

Hiebert, J. & Carpenter, Thomas P. (1992) Learning and Teaching with Understanding. In D. Grouws (Ed.) *Handbook of Research on Mathematics Teaching and Learning,* NY: MacMillan (65-97).

Hiebert, J., Carpenter, T. P., Fennema, E., Fuson, K., Human, T., Murray, H., Olivier, A., & Wearne D. (1996). Problem solving as a basis for reform in curriculum and instruction: The case of mathematics. *Educational Researcher, 25*(4), 12-21.

Hiebert, J., Carpenter, T. P., Fennema E., Fuson K., Human P., Murray H., Olivier, A., & Wearne D., (1997). Making Mathematics Problematic: A Rejoinder to Prawat and Smith. *Educational Researcher, 26*(2), 24-26.

Hiebert, J. & Wearne, D. (1986). Procedures over concepts: The acquisition of decimal number knowledge. In J. Hiebert (Ed.), *Conceptual and procedural knowledge: The case of mathematics* (pp. 199-223). Hillsdale, NJ: Lawrence Erlbaum.

Hiebert, J. & Wearne, D. (1992). Links between teaching and learning place value with understanding in the first grade. *Journal for Research in Mathematics Education, 23*(2), 98-122.

Hiebert, J., & Wearne, D. (1993). Instructional tasks, classroom discourse, and students' learning in second-grade arithmetic. *American Educational Research, 30*(2), 393-425.

Hiebert, J. & Wearne, D. (1996). Instruction, understanding, and skill in multidigit addition and subtraction. *Cognition and Instruction, 14*(3), 251-283.

Hoyles, C. (1986). Scaling a mountain: a study of the use, discrimination and generalisation of some mathematical concepts in LOGO environment. *European Journal of Psychology of Education, 1*(2), 111-126.

Jardine, D. W. (1994). On the ecologies of mathematical language and the rhythms of the earth. In P. Ernest (Ed.) *Mathematics, education, and philosophy: An international perspective* (pp. 109-23), London, England: Falmer.

Jardine, D. W. (1999). The real work is what we really do. In M. Morris, M. Aswell Doll, & W. F. Pinar (Eds.). *How we work* (pp. 33-40). New York: Peter Lang
Jardine, D. W. (2000). *Under the tough old stars: Ecopedagogical essays.* Brandon, VT: The Foundation for Educational Renewal.
Kabat-Zinn, J. (1994). *Wherever you go, there you are.* New York, NY: Hyperion.
Kieren, T. E. (1990). Understanding for the teaching of understanding. *The Alberta Journal of Educational Research, 36*(3), 191-201.
Kieren, T. E., & Pirie, S. E. B. (1991). Recursion and the mathematical experience. In L. Steffe (Ed.) *Epistemological foundations of mathematical experience* (pp. 78-101). New York, NY: Springer-Verlag.
Kieren, T. E., Pirie, S. E. B., & Gordon Calvert, L. (1999). Growing minds, growing mathematical understanding: Mathematical understanding, abstraction and interaction. In L. Burton (Ed.), *Learning mathematics, from hierarchies to networks* (pp. 209-231). London, England: Falmer.
Kieren, T. E., & Pirie, S. E. B. (1992). The answer determines the question– interventions and the growth of mathematical understanding. In W. Geeslin & K. Graham (Eds.), *Proceedings of the 16th Annual Meeting of the International Group for the Psychology of Mathematics Education* (Vol. 2, pp. 1-8). Durham, NH: Program Committee.
Kieren, T. E., Gordon Calvert, L., Reid, D., & Simmt, E. (1995). Co-emergence: Four portraits of mathematical activity. Paper presented at the Annual Meeting of the American Educational Research Association, San Francisco, CA.
Kieren, T., & Simmt, E. (2002). Fractal filaments: A simile for observing collective mathematical understanding. In D. Mewborn, P. Sztajn, D. White, H. Wiegel, R. Bryant, & K. Nooney (Eds.), *Proceedings of the 24th Annual Meeting of the North American Chapter of the International Group for the Pshychology of Matheamtics Educatio,* (Vol. 2, pp. 865-874). Columbus, OH: Eric Clearinghouse for Science, Mathematics, and Environmental Education.
Kieren, T. E., Simmt, E., Gordon Calvert, L., & Reid, D. (1996). Occasioning in mathematical cognition. Paper presented at the Annual Meeting of the American Educational Research Association, New York, NY.
Korzybski, A. (1931). A non-Aristotelian system and its necessity for rigour in mathematics and physics. A paper presented before the American Mathematical Society at the New Orleans, Louisiana, meeting of the American Association for the Advancement of Science, December 28, 1931. Reprinted in Science and Sanity, 1933, p. 747–61.
Kotagama, S. (1993) Practice what you preach: A lesson for sustainable living. In Pablo Piacentini (Ed.) *Story earth: Native voices on the environment.* (pp. 113-121), San Francisco, CA: Mercury House.
Koyré, A. (1958). *From the closed world to the infinite universe.* New York, NY: Harper & Brothers.
Kruteskii, V. A. (1976). *The psychology of mathematical abilities in schoolchildren,* J. Kilpatrick & I. Wirszup, (Eds.). Chicago: University of Chicago.
Lakoff, G., & Johnson, M. (1980). *Metaphors we live by.* Chicago: University of Chicago.
Lakoff, G., & Johnson, M. (1999). *Philosophy in the flesh: The embodied mind and its challenge to western thought.* New York: Basic Books.
Lakoff, G. & Núñez, R. E. (2000). *Where mathematics comes from: How the embodied mind brings mathematics into being.* New York: Basic Books.
Lesh, R. (1981). Applied mathematical problem solving. *Educational Studies in Mathematics, 12*(2), 235-264.
Lesh, R., Landau, M. & Hamilton, E. (1983). Conceptual models and applied mathematical problem-solving research. *In M. Landau and R. Lesh (Eds.) Acquisition of mathematical concepts and processes.* New York: Academic.
Lesh, R., Post, T., & Behr, M. (1987). Representations and translations among representations in mathematics learning and problem solving. In C. Janvier (Ed.), *Problems of representation in the teaching and learning of mathematics,* (pp. 33-40). Hillsdale, NJ: Lawrence Erlbaum.
Lewontin, R. (1983). The organism as the subject and object of evolution. *Scientia, 118,* 63-82.
Liedtke, W. W. (1983). *It's mathematics 3 tests.* Toronto, Ont: Gage.
Lindquist, M., Carpenter, T. P., Silver, E. A., Matthews, W., (1983). The third national mathematics assessment: Results and implications for elementary and middle schools. *Arithmetic Teacher,*

31,(4), 14-19.

Lindsay, P. H., & Norman, D. A., (1984). Procesy przetwarzania informacji u cztowieka, Wprowadzenie do psychologii. Polish translations of *Human information processing: An introduction to psychology.* Warszawa: PWN.

Locke, J. (1985) An essay concerning human understanding. In I. Gogut-Subczynska (Ed.), *Modern philosophical thought in Great Britain: Part I,* (pp., 32-39). Wydawnictwa: Uniwersytetu Warszawskiego.

Martin, L. C., & Towers, J. (2009). Improvisational coactions and the growth of collective mathematical understanding. *Research in Mathematics Education, 11*(1), 1-20.

Martin, L. C., & Towers, J. (2010). Comparing interaction and improvisational coaction in mathematical collaboration. In Brosnan, P., Erchick, D. B., & Flevares, L., (Eds.), *Proceedings of the 32nd Annual Meeting of the North American Chapter of the International Group for the Psychology of Mathematics Education* (Vol. 6, pp. 386-391). Columbus, OH: The Ohio State University.

Maturana, H. (1987). Everything is said by an observer. In W. I., Thompson (Ed.). *Gaia: A way of knowing* (pp. 65-82). New York: Lindisfarne.

Maturana, H. R. (1988a). *Ontology of observing: The biological foundations of self consciousness and the physical domain of existence.* Conference Workbook: Texts in Cybernetics, American Society for Cybernetics Conference, (parts 1-9). Felton, CA.

Maturana, H. R. (1988b). Reality: The search for objectivity or the quest for a compelling argument. *The Irish Journal of Psychology, 9*(1), 25-82.

Maturana, H. R. (1991). Science and daily life: The ontology of scientific explanations. In F. Stïer (Ed.). *Research and reflexivity.* (pp. 30-52). London: Sage.

Maturana, H. R. (1995). The nature of time. [On-line] http://www.inteco.cl/biology/nature.htm.

Maturana, H. R. (1997a). Ask Dr. Maturana. [On-line] http://www.inteco.cl/biology/ask9797-1.htm.

Maturana, H. R. (1997b). Metadesign. [On-line] http://www.inteco.cl/articulos/metadesign.htm.

Maturana, H. R. (1998). Metadesign. [On-line] http://www.inteco.cl/articulos/metadesign_parte3.htm.

Maturana, H. (2001, May). Our genome does not determine us. In P. Bunnell (Chair), *Remaining human.* Public Forum conducted at the meeting of the American Society for Cybernetics, Vancouver, BC.

Maturana, H., & Varela, F. (1980). Autopoiesis and cognition. Philosophy of Science, 42, xxi.

Maturana, H., & Varela, F. (1987). *The tree of knowledge: The biological roots of human understanding.* Boston: Shambhala.

McLuhan, T. C. (1987). *Touch the Earth.* New York: Harper & Row.

Merleau-Ponty, M. (1968). *The visible and the invisible.* A. Lingis (Trans.). Evanston, ILL: Northwestern University.

Minsky, M. & Papert, S. (1972). *Research at the laboratory in vision, language, and other problems of intelligence.* Cambridge, MA: Artificial Intelligence Laboratory, MIT.

Ministry of Education (Province of British Columbia) (2007). *Mathematics K to 7: Integrated resource package 2007.* Victoria, BC, Canada: Ministry of Education.

Naess, A. (1985). The shallow and the deep, long-range ecology movement: A summary. In B. Devall & G. Sessions (Eds.). *Deep ecology: Living as if nature mattered* (pp. 95-100). Salt Lake City, UT: Peregrin Smith.

Naess, A. (1986). Self-realization: An ecological approach to being in the world. The Fourth Keith Roby Memorial Lecture in Community Science, Murdoch University, Western Australia.

Naess, A. (1988). Deep ecology and ultimate premises. *The Ecologist, 18*(4-5), 128-131.

Naess, A. (1996). Living a life that reflects evolutionary insight. *Conservation Biology, 10*(6), 1557-1559.

Naess, A. (2008). *The ecology of wisdom: Writings by Arne Naess* (A. Drengson & B. Devall, Eds.), Berkeley, CA: Counterpoint.

National Council of Teachers of Mathematics, (2000). *Principles and standards for school mathematics.* Reston, VA: The Council.

National Film Board of Canada (NFB). (1961). *Dance squared* [Film]. Montreal, QC, Canada: The Board.

National Film Board of Canada (NFB). (1969). *Notes on a triangle* [Film]. Montreal, QC, Canada: The Board.

New, R. (1993). Cultural variations on developmentally appropriate practice: challenges to theory and practice. In C. Edwards, L. Gandini, & G. Forman (Eds.) *The hundred languages of children*. (pp. 215-234), Norwood, NJ: Ablex.

Orr, D. W. (1992). *Ecological literacy: Education and the transition to a postmodern world.* New York: State University of New York.

Orr, D. (1994). *Earth in mind: On education, environment, and the human prospect.* Washington, DC, Covelo, CA: Island.

Pegg, J. (1997). Interpreting the demands of geometry questions within an extended form of the van Hiele theory. In N. Scott & H. Hollingsworth (Eds.), *Mathematics: Creating the future. Proceedings of the 16th Biennial Conference of the Australian Association of Mathematics Teachers* (pp. 241-246). Adelaide: AAMT.

Pegg, J. (1998). Broadening the descriptors of van Hieles' Levels 2 and 3. In F. Biddulp & K. Carr (Eds.), People in mathematics education. *Proceedings of the 22nd Annual Conference of the Mathematics Education Research Group of Australasia* (pp. 391-399). Rotorua, New Zealand: MERGA.

Pegg, J. & Currie, P. (1998). Widening the interpretation of van Hiele's Levels 2 and 3. In A. Olivier & K. Newstead (Eds.), *Proceedings of the 22nd Annual Meeting of the International Group for the Psychology of Mathematics Education* (Vol. 3, pp. 335-342). Stellenbosch, South Africa: PME.

Pegg, J. & Davey, G. (1998). Interpreting students' understanding of geometry: A synthesis of two models. In R. Lehrer & D. Chazan (Eds.), *Designing learning environments for developing understanding of geometry and space*, (pp. 109-135). Hillsdale, NJ: Erlbaum.

Piaget, J. (1977). Mental images. In H. E. Gruber & J. J. Voneche (Eds.), *The essential Piaget* (pp. 652-684). New York: Basic Books.

Piaget, J. (1985). *The equilibration of cognitive structures* (T. Brown and K. J. Thampy, trans.), Cambridge, MA: Harvard University.

Pimm, D. (1985). *Symbols and meanings in school mathematics.* London: Routledge.

Pimm, D. (1987). *Speaking mathematically: Communications in mathematics classrooms.* London: Routledge & Kegan Paul.

Pirie, S. E. B., & Kieren, T. (1989). A recursive theory of mathematical understanding. *For the Learning of Mathematics, 9*(4), 7-11.

Pirie, S. E. B., & Kieren, T. (1994a). Beyond metaphor: Formalising in mathematical understanding within constructivist environments. *For the Learning of Mathematics, 14*(1), 39-43.

Pirie, S. E. B., & Kieren, T. (1994b). Growth in mathematical understanding: How can we characterise it and how can we represent it? *Educational Studies in Mathematics, 26*(2-3), 165-190.

Postman, N. (1995). *The end of education: Redefining the value of school.* New York: Knopf.

Presmeg, N. C. (1986a). Visualization and mathematical giftedness. *Educational Studies in Mathematics, 17(3),* 291-311.

Presmeg, N. C. (1986b). Visualization in high school mathematics. *For the Learning of Mathematics, 6*(3), 42-46.

Reddy, M. (1978). The conduit metaphor: A case of frame conflict in our language about language. In A. Ortony (Ed.), *Metaphor and thought* (2nd ed., pp. 164-201). Cambridge, UK: Cambridge University.

Reiulle Irons, R. & Irons, C. J. (1989). Language experiences: A base for problem solving. In P. R. Trafton & A. P. Shulte (Eds.), *New directions for elementary school mathematics* (pp. 85-98). Reston, VA: The Council.

Ricoeur, P. (1989) Jezyk, tekst, interpretacja (Polish translation of: 1. *Interpretation thoery: Discourse and the surplus of meaning*; 2. La tâche de l'hermeneutique, 3. La fonction hermeneutique de la distanciation; 4. La mètaphore et le problème central de l'hermeneutique; 5. L'appropriation) Warszawa: PIW.

Rinaldi, C. (1990). Social constructivism in Reggio Emilia, Italy. Keynote Address, annual conference of the Association of Constructivist Teachers, (B. Rankin and L. Gandini, Trans.). Northampton, MA.

Roszak, T. (1994). *The cult of information*. Berkeley, CA: University of California.

Ryle, G. (1949). The concept of mind. London, England: Hutchinson.

Schoenfeld, A. H. (1985). *Mathematical problem solving*. San Diego, CA: Academic.

Schoenfeld, A. H. (1987). What's all the fuss about metacognition? In A. Schoenfeld (Ed.), *Cognitive science and mathematics education* (pp. 189-215). Hillsdale, NJ: Lawrence Erlbaum.

Schoenfeld, A. H. (1989a). Exploring the process problem space: Notes on the description and analysis of mathematical processes. In C. Maher, G. Goldin, & R. B. Davis (Eds.), *Proceedings of 11th Annual Meeting of the North American Chapter of the International Group for the Psychology of Mathematics* Education (pp. 95-113). New Brunswick, NJ: Rutgers Centre for Mathematics, Science, and Computer Education.

Schoenfeld, A. H. (1989b). Teaching mathematical thinking and problem solving. In L. B. Resnick & B. L. Klopfer (Eds.), *Toward the thinking curriculum: Current cognitive research* (pp. 83-103) (1989 Yearbook of the American Society for CurriculumDevelopment). Washington, DC: ASCD.

Schoenfeld A. H. (1991) On Mathematics as sense-making: An informal attack on the unfortunate divorce of formal and informal mathematics. In J. F. Voss, D. N. Perkins, & J. W. Segal (Eds.) *Informal Reasoning and Education* (pp. 311-343). Hillsdale, HJ: Lawrence Erlbaum.

Schoenfeld, A. H. (1992). Learning to think mathematically: Problem solving, metacognition, and sense making in mathematics. In Douglas A. Grouws (Ed.) *Handbook of research on mathematics teaching and learning: A project of the national council of teachers of mathematics* (pp. 334-370). New York: Macmillan.

Schoenfeld, A. H. (1996). In fostering communities of inquiry, must if matter that the teachers knows the "answer"? *For the Learning of Mathematics, 16*(3), 11-16.

Scieszka, J. & Smith, L. (1995). *Math Curse*. New York: Viking.

Sfard, A. (1991). On the dual nature of mathematical conceptions: Reflections on processes and objects as different sides of the same coin. *Educational Studies in Mathematics, 22*(1), 1-36.

Sfard, A. (1992). Operational origins of mathematical objects and the quandary of reification--The case of function. In E. Dubinsky & G. Harel (Eds.), *The concept of function: Aspects of epistemology and pedagogy* (MAA Notes 25, pp. 59-84). Mathematics Association of America.

Sfard, A. (1994). Reification as the birth of metaphor. *For the Learning of Mathematics, 14*(1), 44-55.

Sfard, A. (1997). Commentary: On metaphorical roots of conceptual growth. In L. English (Ed.), *Mathematical reasoning: Analogies, metaphors, and images* (pp. 339-372). Mahwah, NJ: Erlbaum.

Sfard, A. (1998) On two metaphors for learning and the dangers of choosing just one. *Educational Researcher, 27(2)*, 4-13.

Sfard, A. (2000). Steering (dis)course between metaphors and rigor: Using focal analysis to investigate an emergence of mathematical objects. *Journal for Research in Mathematics Education, 31*(3), 296-327.

Sfard, A. (2007). When the rules of discourse change, but nobody tells you: Making sense of mathematics learning from a commognitive standpoint. *Journal of Learning Sciences, 16*(4), 567-615.

Sfard, A. & Kieran, C. (2001). Cognition as communication: Rethinking learning-by-talking through multi-faceted analysis of students' mathematical interactions. *Mind, Culture, and Activity, 8(1)*, 42-76.

Sierpinska, A. (1987). Humanities students and epistemological obstacles related to limits. *Educational Studies in Mathematics, 18*(4), 371-397.

Sierpinska, A. (1990). Some remarks on understanding in mathematics. *For the Learning of Mathematics, 10*(3), p. 24-36.

Sierpinska, A. (1993). On the development of concepts according to Vygotski. *FOCUS on Learning Problems in Mathematics, 15*, p. 2-3.

Sierpinska, A. (1994). *Studies in mathematics education series: 2: Understanding in mathematics.* London, England: Falmer.

Sierpinska, A. & Viwegier, M. (1989). How & when attitudes towards mathematics and infinity become constituted into obstacles in students. In G. Vergnaud, J. Rogalski, & M. Artigue (Eds.), *Proceedings of the 13th Annual Meeting of the International Group for the Psychology of Mathematics* Education (Vol. 3, pp. 166-173). Paris, France: G. R. Didatique, CNRS.

Silver, E. A. (1994). On mathematical problem posing. *For the Learning of Mathematics, 14*(1), 19-27.

Simmt, E. S. M. (2000). *Mathematics knowing in action: A fully embodied interpretation.* Unpublished doctoral dissertation, University of Alberta, Edmonton, AB, Canada.

Skemp, R. (1978). Relational and instrumental understanding. *Arithmetic Teacher, 26(3)*, 9-15.

Skemp, R. (1979) Goals of learning and qualities of understanding. *Mathematics Teaching, 1979*(88), 44-49.

Smucker, B. & Wilson, J. (1995). *Selina and the bear paw quilt*. Toronto, ON: Lester.

Snyder, G. (1990). *The practice of the wild.* San Francisco: North Point.

Steffe, L. P., & Tzur, R. (1994). Interaction and children's mathematics. In P. Ernest (Ed.), *Constructing mathematical knowledge: Epistemology in mathematical education* (pp. 8-33). London, England: Falmer.

Stigler, J. W. & Hiebert, J. (1997). Understanding and improving classroom mathematics instruction: An overview of the TIMSS video study. *Phi Delta Kappan, 79*(1), 14-21.

Suzuki, D. (1997). *The sacred balance: rediscovering our place in nature.* Vancouver, BC & Toronto, ON, Canada: Greystone Books.

Suzuki, D. & Dressel, H. (1999). *From naked ape to superspecies: A personal perspective on humanity and the global eco-crisis.* Toronto, ON, Canada: Stoddart.

Suzuki, D. & Taylor, D. (2009). *The big picture: Reflections on science, humanity, and a quicly changing planet.* Vancouver, BC, Canada: Greystone Books.

Tall, D. (1991). Intuition and rigor: The role of visualization in the calculus. In W.Zimmerman & S. Cunningham (Eds.), *Visualization in teaching and learning mathematics* (pp. 105-120). MAA Notes #19.

Taplin, M. (1995). An exploration of persevering students' management of problem solving strategies. *Focus on Learning Problems in Mathematics, 17*(1), 49-61.

Thom, J. S. & Pirie, S. E. B. (2001). Thinking through ecological metaphors in mathematics education. In S. Gunn & A. Begg (Eds.), *Mind, Body, and Society: Emerging understandings of knowing and learning* (pp. 45-52). Melbourne, Australia: University of Melbourne.

Thom, J. S. & Pirie, S. E. B. (2002). Problems, perseverance, and mathematical residue. *Educational Studies in Mathematics, 50*(1), 1-28.

Towers, J., Martin, L.C., & Pirie, S.E.B. (2000) Growing mathematical understanding: Layered observations. In Fernandez, M.L. (Ed.) In M. L. Fernandez (Ed.), *Proceedings of the 22nd Annual Meeting of the North American Chapter for the International Group of the Psychology of Mathematics Education* (Vol. 1, pp. 225-230). Tucson, AR: ERIC.

Towers, J., Martin, L., & Pirie, S. E. B. (2006). Collective mathematical understanding as improvisation. *Mathematical Thinking and Learning 8*(2) p. 149-183.

Uhrmacher, P. B. (1997). The curriculum shadow. *Curriculum Inquiry, 27*(3), 317-329.

Usiskin, Z. (1996). Mathematics as a language. In P. C. Elliot (Ed.), *Communications in mathematics: K-12 and beyond* (pp. 231-243). Reston, VA: NCTM.

van Hiele, P. M. (1986). *Structure and insight.* Orlando, FL: Academic.

van Hiele-Geldof, D. (1984). Didactics of geometry as a learning process for adults. In D. Fuys, D. Geddes, & R. Tischler (Eds.), *English translation of selected writings of Dina van Hiele-Geldof and Pierre M. van Hiele* (pp. 217-233). Brooklyn, MA: Brooklyn College. (Original work published in 1958).

van Manen, M. (1986). *The tone of teaching.* Richmond Hill, ONT: Scholastic.

van Manen, M. (1991). *The tact of teaching: The meaning of pedagogical thoughtfulness.* London, ON, Canada: Althouse.

Varela, F. (1999) *Ethical know-how: action, wisdom, and cognition.* Stanford, CA: Stanford University.

Varela, F. J., Thompson, E., & Rosch, E. (1996). *The embodied mind: Cognitive science and human experience* (2nd ed.). Cambridge, MA: MIT.

Vinner, S. (1989). The avoidance of visual consideration in calculus students. *Focus of Learning Problems in Mathematics, 11, (1-2),* 149-156.

Vygotsky, L. S. (1987). Thinking and speech. In R. W. Rieber and A. S. Carton (Eds.), *The collected works of L. S. Vygotsky* (N. Minick, Trans.). New York: Plenum.

Walter, M. I., & Brown, S. I. (1993). *Problem posing: Reflections and applications.* Hillsdale, NJ: Lawrence Erlbaum.

Wheeler, D. H. (ed.) (1967). *Notes on Mathematics in Primary Schools.* Cambridge, MA: Cambridge University.

Worster, D. (1990). *Nature's Economy: A History of Ecological Ideas.* Cambridge: Cambridge University.

Yackel, E., & Cobb, P. (1996). Sociomathematical norms, argumentation, and autonomy in mathematics. *Journal for Research in Mathematics Education,* 27(4), 458-477.

Zazkis, R., Dubinsky, E., & Dautermann, J. (1996). Coordinating visual and analytic strategies: a study of students' understanding of the group D4. *Journal for Research in Mathematics Education, 27*(4), 435-457.

INDEX

Abram, David, 8, 61, 183, 341, 368, 370, 372, 373, 375
Aoki, Ted, 370
actions, identities, and thinking: anthropocentric, exploitive, competitive, 2;
back to basics, 53;
collective, 21, 22; commoditized, 2, 32; connected, intuitive, 2; cooperative, qualitative, and conservative, 2-3; cultural, 24-29;
eco-centric, 2-3; about ecological forms of mathematics, 356; ecological-systemic, 11, 13-29, 31, 45, 362, 363, 369; embodied, 170, 364; about environment, 2-3, 8, 141-142;
holistic, 2, 3, 35, 170;
integrative, 2, 3, 35;
mechanistic, linear, rational, reductionist, 2-3; metaphorical nature of, 27, 364;
nonlinear/nonlinearity of, 170;
taken for granted, 364, 365
addition, 204, 205, 311: repeated, 219, 220, 221, 258, 262, 263, 287, 288, 290, 296, 297, 298, 299, 300, 301, 319, 343
AIMS (Activities Integrating Mathematics and Science), 374
application of the SOLO taxonomy to the analysis of students' geometric understanding, 94-95
are you a square or an oblong?, 323-339
arisings, 157
Aristotle, 372
atmosphere, 163-165
atmosphere and re-cognizing differences, 163-171
attending to the physical space in the classroom, 175-180
autopoiesis, 16
autonomous: child, 26, 64, 65, 244
average (mode), 282
Bachelard, Gaston, 97
back on the strip: my university years, 41
Baroody, Arthur, 370, 371
Basso, Keith, 34, 35, 51, 369
Bateson, Gregory, 27, 140, 237, 239, 304, 341, 369, 370, 371, 374, 375
Bateson, M. Catherine, 10, 11, 31, 40, 44, 45, 61, 206, 362, 368, 369, 370, 373, 374, 375, 376
Behr, Merlyn, 83
being: autopoietic, 16, 120, 128; collective ways of, 16; cultural ways of, 16, 24-29; human, 16; individual ways of, 15-19, 26, 29; social ways of, 16, 20-23, 29
Bergeron, Anne, 90, 91
Bergeron, Jacques, 90, 91, 96
Berliner, Paul, 133
Berry, Wendell, 51, 370
Beth, Evert, 103
Biggs, John, 92-95
Bishop, Alan, 110
Blades, David, 367
Blake, Anthony, 92
Boukhssimi, Driss, 90
Bowers, Chet, 25, 27, 166-167, 369, 372
Bowers, Janet, 127, 132
British Columbia Association for Mathematics Teachers (BCMAT), 373
Brown, Stephen, 373
Brownell, William, 83
building bridges to connect informal and symbolic mathematics with student understanding, 84-85
Bunnell, Pille, 5, 368
Burkhardt, Hugh, 349, 375

Camus, Albert, 369
can you build my design?, 309-310
Capra, Fritjof, 2, 25, 27, 61, 368, 369, 370
Carpenter, Thomas, 83, 84
Carraher, David, 84, 117
Carraher, Terezinha, 84, 117
celebrating together, 312-320

change: eco-centric, 2-3; in the mathematics classroom, 2-3, 35-36; (mind-)spaces, 52-53, 54-55, 143-144; structural, 16, 17, 18, 21; in teaching mathematics, 11, 31, 45, 54-55, 166-171, 173-366; teaching as practice to praxis, 55-56, 167-171, 203, 363
classroom mathematics:
ecological metaphors for, 140-142, 158-160, 200, 235; ecological view of, 8-10, 13-29, 150-160, 363; emergent, 147-161;
individual within collective, 127-128, 132, 234, 266, 267, 268, 365, 370; integrative, 171, 267, 335, 370;
mechanistic metaphors of, 52-53, 64-66, 138-139, 158-160, 200;
reflections revealing the fragility of, 37-44; reimagining, 53, 144, 200; as residue, 86, 142, 149, 155, 156, 159-160, 171, 365, 371;
as source, 142, 149, 156, 157, 159-160, 171, 205, 365
Clements, Ken, 110
Cobb, Paul, 127, 132
co-emergence, 170: curricula and understanding, 147-161; organism and environment, 18-19; social phenomena, 21-23; teaching and learning, 69, 168, 203; teaching and research, 55-56. *See also* mathematical understanding
co-evolution, 170:
mathematics class and classroom environment, 141; mathematics content and mathematical processes, 148-149;
organism and environment, 18-19;
teaching and learning, 18-19, 69, 168, 203; teaching and research, 55-56; theory and practice, 364
cognition: all doing is knowing, all knowing is doing, 67, 125, 127, 130-131, 134, 170; inseparability of communication and, 127; as a process of living, 16, 17, 19, 21, 22, 25; re-cognize, 376
coherence: of metaphorical manners, thinking, and conceptions of teaching, 67-70, 71-72, 137-142, 167-171, 202, 349
Collis, Kevin, 92-95
common space, 10, 40, 44
complex, 138-139, 206-208, 363
complex circularity, 170: of cognition, 369; learners and environment, 18-19; mathematics and mathematical processes, 142, 148-149; organism and environment, 18-19; teaching and research, 55-56; theory and practice, 364
complexization, 101
complicated, 138-139, 206-208
conceptualizations of time, 142, 181-198, 200, 202, 205, 206-208, 235, 364
conceptual thinking as synthesis, 101
connecting mathematical processes of knowing with objects of knowledge: operational and structural conceptions of mathematics, 120-121
connecting as shape-shifting, 269-284
connecting me, connecting us, 323-339
connecting us to it, 341-345
construction of mathematical understanding, 86-87
Cottrill, James, 87, 101, 102, 105
creating patterns which connect, 237-302
Cristaldi, Kathryn, 372
Cult of Information, 155
culture: Chinese, 26; cognitive circles, 25, 26, 33; Japanese, 26; map, 27, 33, 369; Western, 26, 27, 368
cultural: behaviour, 24-29; beliefs, 26-27; map, 27, 369; ways of being, 24-29
curriculum:
British Columbia, 66, 203, 370, 373;
commoditized, 50-51, 52-53, 143-144;
jigsaw puzzle, 66, 138-139;
lived/living, 171, 202, 203, 235, 370;
map, 140, 171, 200-205, 365; mechanistic, 50-51, 52-53, 66, 143-144; metaphors of, 50-51, 52-53, 66, 140, 143-144, 235;
objectified, 52-53, 143-144;
scripted as unscripted, 147-150, 267, 365; shadow, 370
Currie, Penelope, 92-95

daimon, 27
Dance Squared, 231, 374, 375
Dautermann, Jennie, 108
Davey, Geoff, 94, 95
da Vinci, Leonardo, 324
Davis, Brent, 18, 46, 130, 132, 138, 369, 370, 371, 372
Davis, Robert, 83, 86-88, 89-90, 112, 121, 142, 371
data: analysis, 269-284, 327, 328, 329-335, 337-339; collecting, 325-328, 336
deeper yet, 71-72

Devlin, Keith, 371
Dewey, John, 98
difference(s): re-cognizing, 168-171; systemic, 141, 168-171; that makes a difference, 169-171
Dionne, Jean, 90
distributive law over addition, 298
diversity, 170: structural, 16, 17
division: 150, 154, 156, 220, 221, 287, 290, 291, 292, 296, 297, 298, 299, 349-350; groups of, 245-264, 320
do the metaphors with which we describe beginnings and ends really matter?, 158-160
does the way one portrays mathematical understanding matter?, 77
Doll, Jr., William, 373
Dressel, Holly, 368
Dreyfus, Tommy, 109
drift: 21, 207; cultural, 25, 369; natural, 19, 369
Dubinsky, Ed, 87, 96, 101-104, 108
Durkin, Kevin, 88

earth: 1, 27, 29; sensuous, 8
ecological: (in)coherence, 3, 10, 35, 51, 54, 68, 69-71, 137, 138-142, 167, 168, 349, 365; mindfulness, 9-10, 165, 170, 363; mindfulness and mathematics education, 6-10, 165, 235
ecology, 8, 368; encircling teaching within, 56; oikos, 8, 9, 141, 167, 168, 169, 363; oecologie, 8; shallow and deep, 368; three faces of, 11, 31, 45, 362
Edwards, Carolyn, 370, 372
Eisenberg, Theodore, 109
embedding and rooting an ecological sense of place for mathematics in the classroom, 173-366
enacting re-conceptualizations of time, 199-235
environment(al): crisis, 1-3; external, 18-19, 64, 65; as living system, 18-19, 181; mathematical, 11, 18, 19, 31, 45, 362;
even numbers, 147, 151-155, 157, 159-160, 267, 271, 289, 291, 292, 319, 343
Even Steven and Odd Todd, 147, 267, 372
Evernden, Neil, 3, 368
evolution: natural drift, 19
examining mathematical understanding from a constructivist perspective, 80-106

Fibonacci, Mrs., 48, 370
Flinders, David, 166-167, 372
Forman, George, 370, 372
four Rs of recursion, 206-208
Fox, Warwick, 368
fractal(s): 214-221
fractions, 154, 318, 350-351, 353-357
free ideas, 266, 267, 341
free ideas and connecting as shape-shifting, 265-284
from the side of pedagogy, 42
Fulghum, Robert, 347, 359, 375

Gadamer, Hans-Georg, 125, 375
Gandini, Lella, 370, 372
geometric understanding: multistructural, 92, 93, 95; Pegg's adapted Van Hiele theory, 94; prestructural, 92, 93, 95; van Hiele model of geometric thought, 94
geometry: 204, 205, 214-217, 219-221, 224-229, 231-234, 309-310, 313-317, 318, 323-339, 342-345; designs, 309-310, 313-317; and positional language, 309-310; spatial patterns, 214-221, 228-229, 231-234
Goldenberg, E. Paul, 110
Gollwitzer, Herman, 109
Gonzales, Nancy, 373
Gorden Calvert, Lynn, 129-130, 368, 369, 375
Gravemeijer, Koeno, 127
Greenman, Jim, 181, 372
growth: in teaching mathematics, 128-131, 170, 203; *See also* mathematical understanding

Haeckel, Ernst, 8
Hallam, Roy, 92
Hamilton, Eric, 96
Hart, Lynn, 112, 117
Heidegger, Martin, 60, 370, 372, 374
Hèraurd, Bernard, 90
Herscovics, Nicolas, 90, 91, 96
Hiebert, James, 32, 33, 83-86, 111, 117, 363, 364, 368, 369, 370, 375, 376
help wanted, 11, 31, 45, 362
Hillman, James, 57
historical transformation, 25
homonymy, 88
homophony, 88
Hoyles, Cecelia, 98

individual ways of being, 15-19, 26, 29
interaction: beyond spaces of, 321-345; significance of, 6-10; social, 21-23; social mathematical, 21-23
interactional spaces for mathematics in the classroom, 243-302
insight, 49, 244, 265

interdependence, 170: classroom mathematics, ecology, and the world, 11, 31, 45, 362, 363; learners and environment, 18-19, 369; mathematics and mathematical processes, 53, 142, 144, 148-149; organism and environment, 18-19, 369; teaching, learning, and mathematics, 63, 168, 171, 349, 366
invisibility: cultural patterns, 25-26, 33; knowing, 33; manners of teaching, 46, 50-51, 52-53, 54-56, 59-70, 143-144, 166; metaphors and metaphorical patterns, 27, 52-53, 60-72, 143-144
Irons, Calvin, 64
in search of mermaids, 352-359
is half of a half really a half?, 350-351

Jardine, David, 38, 57, 61, 322, 323, 369, 370, 375
Johnson, Mark, 61, 370
just hand them down the mathematics... or not?!, 52-53, 54, 143-144

Kabat-Zinn, Jon, 372
keeping our classes connected, 306
Kieran, Carolyn, 125, 126
Kieren, Thomas, 96, 128-132, 368, 369, 370 372, 373, 375
knowing and knowledge:
acquired, 120;
embodied, 170;
formal, 81, 83, 91, 92, 111, 112, 115, 126, 130, 135;
individual and collective, 141, 142, 170, 365; informal, 84, 85. 86, 115, 135; intuitive, 90;
linear, mechanistic, and hierarchical, 2-3, 61, 65-66, 368;
nonlinear and holistic, 2-3, 368;
participatory, 120; place(s) of, 292; as product, producer, and process, 130;
of social systems, 21-23;
tacit, 166;
see also cognition and mathematical understanding
Korzybski, Alfred, 27, 60, 140, 369, 370, 371
Kotagama, Sarath, 368
Koyré, Alexandre, 372
Kruteskii, Vadim, 109

Lakoff, George, 60, 61, 370
Landau, Marsha, 96
language and languaging, 22-28: cultural, 27, 369, 372; mathematical, 142, 149, 155; mathematics classroom, 60-72, 365; metaphorical, 27, 161, 200; social systems, 22-23, 369
languaging emergent mathematics, 143-161
Lesh, Richard, 83, 96, 373
(let it roll) extraordinary equations for you!, 311
Lewin, Philip, 87, 96
Lewontin, Richard, 18-19, 369
liberating constraints, 143-144, 147, 372
Liedtke, Werner, 373
limitations of metaphors: issues of integration, overprojections, and confinement, 121-122
linking written symbols with understanding, 84
Lindquist, Mary, 84
Lindsay, Peter, 96 369
Locke, John, 98
Luce-Kapler, Rebecca, 18, 46, 138, 369, 370, 371, 372

making (three) space(s) for recursion, 209-212
Martin, Lyndon, 130, 132-133
materials matter, 180
Math Curse, 48, 370
mathematical: ambiguities, 87-90; gifts, 209-212, 307-311; language, languaging and residues of learning, 151-155; objects, 102, 120, 125
mathematical concepts, 89-90
mathematical processes:
analytic strategies, 108-110;
coaction, 133; communicating, 53, 64, 142, 144, 148; communication as commognition, 127-128; condensation, 120, 121, 126; connecting, 53, 64, 142, 144, 148; control skills, 116-117; coordination, 104, 109;
empirical abstraction, 103; encapsulation, 104, 109;
identification and discrimination, 99-101; interiorization, 104, 109, 120, 121, 127;
monitoring mathematical actions, 87, 95, 115, 116;

NCTM processes, 53, 64, 142, 144, 148, 371, 372;

perseverance, 116-117; problem solving, 53, 64, 142, 144, 148; pseudo-empirical abstraction, 103;

reasoning and proof, 53, 64, 142, 144, 148; reciprocal nature of mathematics and, 148-149; reflective abstraction, 101, 102-104, 109; reification, 120, 121, 127; representing, 53, 64, 142, 144, 148; reversal, 104, 109;

visual strategies, 108-110;

see also mathematical understanding

mathematical understanding:

abstraction, 101; actions, 101; as actions, processes, objects, and schemas (APOS) via reflective abstraction and genetic decomposition, 101-105; attended focus, 126, 127;

bidirectional, 111;

chain complexes, 99-100; circular, 111; collaborative and social, 111, 127; collective, 127-128, 132-133, 234, 266, 267, 268, 276, 278-279, 365; from within a commognitive framework, 127-128; comparing personal and collective, 127-128, 132, 267, 365; complex circularity, 18-19, 126-127; conceptions of, 71-72; conceptual, 64; as connecting and constructing networks, 83-86; as constructed, 80-106; construction of, 86-87; as conversation, 125-126;

descriptive not prescriptive model, 130-131; discrimination, 98, 99; don't need boundaries, 129, 207; as dynamic, 159, 107-123;

co-emergence of individual and collective understandings 266-268, 283-284, 365, 368, 369, 370, 375.

emphasizing the dynamic nature of, 107-123; empirical abstraction, 103; as enacted, 124-134; from an enactive perspective, 124-134; endorsed narratives, 127; etiquette, 133; extended abstract, 92, 94;

folding back, 129, 130, 263, 285, 287, 302; formal, 64, 81, 83; formalising, 128-129, 131, 288, 298, 302; formalization, 90-91; fractal like, 130-131, 235;

generalization, 98, 104, 109; genetic decomposition, 104-105; as growth, 171, 206: collective, 234, 266, 267, 268, 276, 278-279, 365; individual, 234, 266, 267;

the Pirie-Kieren model, 128-131, 170, 203, 207, 373;

Hiebert's three sites for linking symbols with, 84; historico-empirical approach, 96-97; hypothetical cognitive structures (HCS), 92;

identification, 98; image having, 128-129, 131, 207, 287; image making, 128-129, 131, 207; importance of the group mind, 133; improvisational nature of collective, 132-133; individual and collective, 125-126, 127-128, 132, 142, 159, 170, 234, 266, 267, 268, 283-284, 365; instrumental, 64, 81-82, 83, 111; intended focus, 126, 127; intuitive, 83, 90-91; inventising, 128-129, 131;

jigsaw puzzle, 86;

leveled but non-linear, 130; logical, 81, 83; logico-mathematical abstraction, 90-91; logico-mathematical procedural, 90; logico-physical abstraction, 90-91; logico-physical procedural, 90;

mathematical words, 127; mediators, 127; mental networks, 83-84, 116; as a mental process of identification and discrimination, 98-101; mental schemas, 87, 101, 102, 103, 104, 112, 135; metaphors, 155; model for analyzing students', 112-113; multistructural, 92, 93, 95;

objects, 102, 120; observing, 128-129, 131; operational, 120, 126; operational concept-metaphor, 126; as overcoming epistemological obstacles, 95-99;

participatory, 127-128; as paths, 22, 140-141, 170, 171, 203, 224; as personal and public forms of communication, 125-127; Pirie-Kieren theory, 128-132; potential concepts, 101; potential pathways, 133; prestructural, 92, 93, 95; primitive doing or knowing, 128-130, 131, 207; problem solving for the development of, 114-115; procedural, 64; processes, 102; as a process and an act, 95-96; pronounced focus, 126, 127; property noticing, 128-130, 131, 207, 290; pseudo abstraction, 103; pseudo-concept of generalization, 100; pursuing a better idea, 133;

as reconstructed, 81, 87, 89, 90, 96, 98, 102; recursive, 18, 128, 130, 279, 281, 373; reflective abstraction, 102-104; reflective dimensions, 83; relational, 64, 81, 82, 83, 92, 93, 111, 116; routines, 127;

Schoenfeld's analytic model, 112-114; Sfard's three stages of mathematical conception, 120-121; striking a groove, 133; structural concept-metaphor, 126; as structure, 16-18, 21; structure of memory: access to resources, 115-117; structuring, 128-129, 131; synthesis, 98;
taken as shared, 132; as taxonomy, 92-95; theoretical portraits of, 77-135; thinking as communication, 125-127, 371;
unistructural, 92, 93, 95;
visual and analytic strategies as interrelated qualities of, 108-110
mathematics:
beyond the classroom, 303-320;
commoditization of, 52-53, 143-144, 200-202; conceptions of, 71-72, 73-76, 145, 158-160;
as embodied, 53, 75-76, 144, 145, 205; emergent, 147-161, 200-202, 365, 370;
as hand-me-downs, 52-53, 143-144;
as a language, 88-89;
metaphors of, 52-53, 65, 142, 143-144, 158-160; multiversal, constitutive, 76, 145, 159, 169;
nature of, role in education, 371;
operational and structural conceptions of, 120-122;
as relations, 43, 351, 357; as residue, 86, 142, 149, 155, 156, 159-160, 171, 365, 371;
as shape-shifter, 76, 142, 204, 269-284; as source, 142, 149, 157, 159-160, 171, 365;
universal, absolute, 74-75; unpredictable, 171, 347-359
mathematics class: nested systems, 141
mathematics classroom: as family household, 9; as place, 34-44, 51, 161, 165, 200, 235, 366; systemic view, 11, 19, 13-29, 31, 45, 168, 200, 244, 362
mathematics instruction:
aquisitional vs. participatory, 119;
as genetic decomposition, 87, 104-105;
Herscovics and Bergeron's analytic framework, 90-91;
metaphors for, 61-72; managerial skills, 117;
provocative, 18, 141, 285, 369;
re-conceptualization of, 3, 41, 42, 44; as structural change, 16-21, 129;
mathematics learner: metaphors of, 64-65; re-conceptualization of, 365
mathematics learning:
collective, 21-22, 127-128, 132-133, 234, 266, 267, 268, 276, 278-279, 365; commognitive conflicts, 128; as connecting, 241-264, 269-284, 323-339, 341-345; as conversation, 125;
ecological view of, 266; epistemological obstacles, 93-99;
individual and collective, 142, 170, 234, 266, 267, 268, 283-284, 365; integration, overprojection, confinement of metaphors, 121-122; invocative, 18, 141, 369;
mechanistic forms, 39, 50, 65, 200; metaphors for, 64-65, 141, 170-171; misconceptions, 87, 89-90, 112
nonroutine, 87, 373;
see also mathematical understanding
mathematics teaching:
(collaborative and) emergent, 147-161, 365; conceptions of, 161, 167-171, 363-366;
ecological metaphors of, 140-142, 167-171, 200, 235;
as facilitative, 64, 67; focused on students' overcoming of epistemological obstacles, 95-99;
manners of, 48-51, 52-53, 143-144; mechanistic forms of, 50-51, 67-68, 138-139; mechanistic, linear, and hierarchical metaphors of, 52-53, 64-68, 143-144;
and not knowing, 349;
one's growth in, 167, 171, 363;
taken for granted manners of, 52-53, 143-144, 200, 356, 365;
see also mathematics instruction
Matthews, Westina, 84
Maturana, Humberto, 16, 21, 22, 25, 26, 61, 120, 125, 130, 149, 169, 369, 370, 371, 372, 373, 374
McLuhan, Teri, 372
measurement: linear, 323-339
meeting with mac, 245-264
Merleau-Ponty, Maurice, 374
metaphorical nature: of mathematical understanding and the processes by which understanding occurs, 118-122; of thinking, 27, 60-72

metaphorical mattering of mathematics, 73-76
metaphors: acquisitionist, 118-119, 120; enactions of, 59-71; participation, 118-119, 120; pervasive and unconscious nature of, 27; re-imagining, 71, 156-160, 167-171, 365; shaping thinking and conceptions of teaching, 32-33, 52-53, 60-70, 143-144, 167-171, 365. *See also* actions, identities, and thinking, classroom mathematics, coherence, curriculum, invisibility, language, mathematical understanding, mathematics, mathematics instruction, mathematics learner, mathematics learning, and mathematics teaching
mind-space: anOTHER, 54-56, 161; ecologicial, 140-142, 366, 149-150, 161, 169, 244, 349, 363; furnishings of, 54-55, 69-70, 137-142, 202; linear, mechanistic, 32, 200, 202; shifts in, 32-33, 71-72, 200, 202
Ministry of Education (British Columbia), 370, 373
Minsky, Marvin, 86
Möbius strip, 40, 41, 42, 43, 44, 206, 208, 373
model of intelligence and forms of mathematical understanding, 81-83
Morehouse, Henry, 372
Moser, James, 84
moving thinking spaces and assessing old furniture, 54
multiplication, 205, 219, 220, 221, 262, 263, 286, 287, 290, 291, 297, 300, 311, 342: groups of, 219, 220, 221, 258-259, 260, 261, 262, 296, 320, 342-343; of fractions, 355-357

Naess, Arne, 25, 368, 369
NCTM (National Council of Teachers of Mathematics), 370, 371, 372, 373
National Film Board, 374, 375
Natural Alien, 3
new furniture, 137-142
New, Rebecca, 373, 374
Newton, Isaac, 372
Nichols, Devilyna, 101, 102, 104, 105, 378
Nickerson, Susan, 132
~~linear~~ and ~~hierarchical~~ views of mathematical understanding, 111-117
Norman, Donald, 96
Notes on a Triangle, 205, 224, 374
number: 286-301, 307, 311, 313, 317; comparing, 319; gymnastics, 204, 314-320; mystery, 209-212; patterns, 205, 214-221, 228-229, 231-234, 342-343;

Núñez, Rafael, 60, 370

Oberg, Antoinette, 367
odd numbers, 147, 151-155, 157, 159-160, 207, 267, 271. 289, 291, 292, 319
oecologie, 8
off the trail, 361-366
okay, but where does one start?, 3
oikos, 8, 9, 141, 167, 168, 169, 363
open-closure, 349, 363, 366
opening mathematical spaces of their own, 286-302
organization, 16
Orr, David, 25, 61, 368, 369

path(s) laid down in walking, 140, 170
pattern(s): which connect(s), 19, 239, 304, 341, 370, 375; cultural, 24-27; of difference, 141, 168-171; of mathematics teaching, 3, 46, 365; metaphorical, 27; spatial and numerical, 214-217, 219-221; 228-229, 231-234, 342-345, 355-357; of thinking, acting, and being, 2, 3, 19, 364
Papert, Seymour, 86
Pegg, John, 92-95
perceptions:
attitudes and, 2-3, 368, 369;
Cartesian, Newtonian, 2, 368;
ecological, 2-3, 363, 368;
of place, 34, 35, 51, 161, 163-165, 200, 235;
systemic, 11, 31, 45, 362, 368;
of the world, 2, 11, 31, 45, 362, 368, 369, 375; of the world and the mathematics classroom, 2-3, 13-29
perturbations: 17, 18, 22; human and nonhuman, 17; in learning, 18, 129, 130
Piaget, Jean, 92, 93, 102, 110
Pimm, David, 84, 88-89, 111
Pirie, Susan, 96, 128-133, 367, 368, 369, 370, 371, 372, 373, 375
place: 33;
beginning with, 35-36;
commoditized, linear, disconnected sense of, 39, 50-51, 200, 202, 235;
diamon of a, 57;
ecological sense of, 161, 200, 235, 364;
mathematics', 341; mathematics classroom as, 34-44, 51, 364;
sense of, 34–36, 200, 235, 369; a student's, 39;

worthiness of sense of, 164, 165, 167
See also perceptions
place-making: in the classroom, 174-366; metaphorical, 161, 235; rituals of, 50-51, 71, 165, 202
place value, 154, 292-295
polysemy, 88
Post, Thomas, 83
Postman, Neil, 27, 60, 369
Presmeg, Norma, 109
problemetizing children's learning of mathematics, 85
property noticing, 128-130, 131, 207, 290

question: so what does classroom mathematics have to do with this? answer: everything!, 3

raising questions and questioning the answers: ~~linear~~ time-spaces for mathematics in the classroom, 201-205
recursion, 170:
of mathematics, teaching, and learning, 206-235;
of patterns of experiences, 155;
as reflecting, 206-208; as renewal, 206-208, 213-221; as relations, 141-142, 206-208, 231-234; as relations: when triangles become square, 231-234; as re-viewing, 155, 206-208, 224-229, 342, 365; as revisiting, 218-221, 365;
spaces for, 209-234;
of teaching and learning, 171;
see also drift
Reddy, Michael, 118
reflect on the reflection, 308
reflection(s): 5; as actions, processes, objects, and schemas (APOS) via reflective abstraction and genetic decomposition, reflective abstraction, and reflective dimensions; revealing the fragility of classroom mathematics, 37-44. *See also* mathematical understanding
reflective symmetry, 308, 322
Reggio Emilia, 204, 369, 370
Reid, David, 130
re-imagining: classroom mathematics, 3, 53, 54, 144; mathematics teaching and learning, 3, 53, 144, 161, 167-171
relations: among the classroom and larger communities, 10; defining sense of place, 34-36, 51, 163-165, 177, 180, 200, 235, 364, 365; as meaning or difference, 141, 169, 375; as paths of understanding, 364, 365, 366; of teaching and learning, 141-142, 167-171
(re-)root(ing): changes, 35-36; the learning space, 33, 357, 363; metaphors, 140-142, 200; thinking, 2
Reuille Irons, Rosemary, 64
re-viewing and seeing differently, 224-229
re-cognizing, 169, 376
rethinking mathematical understanding from an enactive from an enactive perspective, 124-134
Ricoeur, Paul, 96
Rinaldi, Carlina, 175, 372
rituals and traditions: anthropocentric and mechanistic, 2-3, 52-53, 143-144; linear, 52-53, 143-144, 202
Rosch, Eleanor, 140, 369
Roszak, Theodore, 155, 372
Ryle, Gilbert, 111

Schliemann, Analucia, 84, 117
Schoenfeld, Alan, 87, 111-117
Scieszka, Jon, 370
scripting an unscripted lesson, 147-150, 365
Selina and the Bear Paw Quilt, 313, 375
settling in, 32-33
settling in(to anOTHER) space, 54-56
Sfard, Anna, 61, 96, 118-122, 125-128, 149, 370
Shape-shifter: mathematics as, 76, 142, 204, 250
shifts in applications, 88
Shire, Beatrice, 88
Sierpinska, Anna, 87, 95-101, 121
significances in the flow of interacting, 5-10
Silver, Edward, 84, 87
Simmt, Elaine, 130, 132
Skemp, Richard, 81-83, 86, 96
skip counting, 246-247, 249, 263, 292
Smith, Lane, 370
Smucker, Barbara, 375
snowflake(s), 204: Koch, 214-217, 218-221, 373; of a different kind, 214-221; taking a closer look at, 342-345
Snyder, Gary, 363, 364, 366, 375, 376
social ways of being, 20-23, 29, 368, 369
space(s): common, 10, 40, 44; ecological thinking, 149; the learning, 33, 357, 363, 365; physical, 175-180; for unpredictable mathematics, 347-359; wanted, 11, 31, 45, 362
Steffe, Les, 369
stepping into Jennifer's classroom, 177-180

Stigler, James, 32, 33, 363, 364, 368, 369, 370, 375, 376
structural coupling, 16-19, 207, 369
structure: evolution of, 16-19;
 mathematical, 16, 17, 18; of a system, 16, 21, 27-28, 169, 369
subtraction: 204, 288, 297, 300; repeated, 219, 245-264, 296, 297, 298, 299, 301; take away one strategy, 306
Sumara, Dennis, 18, 46, 138, 369, 370, 371, 372
surfacing and noticing, 59-70
sustainability, 170: ethic for living, 3
Suzuki, David, 368
systems:
 autopoietic, 16, 120, 128, 369;
 cultural, 25-29;
 living, 171;
 organization of, 16;
 self-making, 16, 368; social, 20-23, 29; social-cultural, 27, 132, 368; structure of, 16

Tall, David, 110
Taplin, Margaret, 117
Taylor, David, 368
teacher: metaphors for, 52-53, 61-72, 141, 143-144; role of, 17-18, 87, 200
teaching:
 collaborative, 303-320; conceptions of, 32, 33, 41, 42, 44, 56, 64-66, 67;
 ecologically responsive, 11, 31, 45, 362, 363; ecological view of, 8, 363;
 furnishings of, 54, 55, 69-70, 137, 138-142, 202
 of mathematics, 86-87; as more than teacher, 141
 praxis, 55, 168, 203, 363;
 and research, 55; re-searching one's, 36, 48, 50, 51, 55, 56, 59-61, 365;
 shaped by one's conception of mathematics, 62, 65, 67, 136;
 taken for granted conceptions of, 365; as taken for granted manners of, 3, 52-53, 61-63, 71, 143-144, 166, 365; turning the soil: the beginning years of, 48-51
text collage: constructivist, 64; ecological, 170
text portrait: constructivist, 106; dynamic, 123; enactive, 134
theoretical portraits of mathematical understanding, 77-135
third time around, 43
Thom, Jennifer, 367, 368, 371
Thompson, Evan, 140, 369
time: 181;
 absolute, 183-184; as all-at-once, 142, 366, 374; assessing conceptions of, 200, 202, 205, 206-208, 235;
 conceptions of, 372;
 and environment, 181;
 linear, 185-197, 200;
 as recursive, 374; re-imagining, 142, 203-204, 209-234, 235;
 time-space, 200, 202, 203, 204, 205, 213, 235, 374;
 whorling, 198
Towers, Jo, 130, 132-133
traditions, tensions, and transitions, 45-72
turning the soil: the beginning years of teaching, 48-51
two and half years later, 235
Tzur, Ron, 369

understandings that speak to one another, 244-264
Urmacher, Bruce, 370
Usiskin, Zalman, 84

vacancy: 11, 31, 45, 362
van Hiele model of geometric thought, 94
van Hiele, Pierre, 94, 96
van Hiele-Geldof, Dina, 94
van Manen, Max, 61, 163-165, 167, 238, 341, 370, 372, 374, 375
Varela, Francisco, 16, 25, 120, 125, 130, 140, 169, 368, 369, 371, 372
Vinner, Shlomo, 86, 89-90, 109, 110, 121
Viwegier, Monika, 100
von Koch, Helge, 373
Vygotsky, Lev, 98, 99, 100, 101

Walter, Marion, 373
we are connected to this earth, 1-3
Wearne, Diana, 83, 84
web of life, 2, 27,
what is a tree?, 3
Wheeler, David, 371
Whitenack, Joy, 127
Wilson, Janet, 375
Wisdom Sits in Places, 35
Worster, Daniel, 368

Yackel, Erna, 127

yes, but we recycle, 2-3

Zazkis, Rina, 108-110